**IFSTA 210
FIRST EDITION**

PRIVATE FIRE PROTECTION AND DETECTION

VALIDATED BY

THE INTERNATIONAL FIRE SERVICE TRAINING ASSOCIATION

PUBLISHED BY

**FIRE PROTECTION PUBLICATIONS •
OKLAHOMA STATE UNIVERSITY**

EDITORIAL STAFF
Jerry W. Laughlin - Editor
Gene P. Carlson - Associate Editor
Charles W. Orton - Assistant Editor
Don Davis - Production Coordinator
Cover photo by Don Davis

PUBLISHED BY **Fire Protection Publications Oklahoma State University**

In cooperation with

The International Association of Fire Fighters
The Insurance Services Office
Numerous State Boards for Vocational Education

Dear Firefighter:

The International Fire Service Training Association (IFSTA) is an organization that exists for the purpose of serving firefighters' training needs. IFSTA is a member of the Joint Council of National Fire Organizations. Fire Protection Publications is the publisher of IFSTA materials. Fire Protection Publications staff members participate in the National Fire Protection Association and the International Society of Fire Service Instructors.

If you need additional information concerning our organization or assistance with manual orders, contact:

Customer Services
Fire Protection Publications
Oklahoma State University
Stillwater, OK 74078-0118
1-(800) 654-4055

For assistance with training materials, recommended material for inclusion in a manual or questions on manual content, contact:

Technical Services
Fire Protection Publications
Oklahoma State University
Stillwater, OK 74078-0118

First Printing, December 1979
Second Printing, February 1981
Third Printing, June 1983
Fourth Printing, November 1987

Fifth Printing, December 1988
Sixth Printing, November 1989
Seventh Printing, November 1990
Eighth Printing, October 1992

Oklahoma State University in compliance with Title VI of the Civil Rights Act of 1964 and Title IX of the Educational Amendments of 1972 (Higher Education Act) does not discriminate on the basis of race, color, national origin or sex in any of its policies, practices or procedures. This provision includes but is not limited to admissions, employment, financial aid and educational services.

© *1979 by the Board of Regents, Oklahoma State University*
All rights reserved
ISBN 0-87939-036-0
Library of Congress 79-556-79
First Edition Published in 1979
Printed in the United States of America

THE INTERNATIONAL FIRE SERVICE TRAINING ASSOCIATION

The International Fire Service Training Association is an educational alliance organized to develop training material for the fire service. The annual meeting of its membership consists of a workshop conference which has several objectives —

> . . . to develop training material for publication
> . . . to validate training material for publication
> . . . to check proposed rough drafts for errors
> . . . to add new techniques and developments
> . . . to delete obsolete and outmoded methods
> . . . to upgrade the fire service through training

This training association was formed in November 1934, when the Western Actuarial Bureau sponsored a conference in Kansas City, Missouri, to determine how all agencies that were interested in publishing fire service training material could coordinate their efforts. Four states were represented at this conference and it was decided that, since the representatives from Oklahoma had done some pioneering in fire training manual development, other interested states should join forces with them. This merger made it possible to develop nationally recognized training material which was broader in scope than material published by an individual state agency. This merger further made possible a reduction in publication costs, since it enabled each state to benefit from the economy of relatively large printing orders. These savings would not be possible if each individual state developed and published its own training material.

From the original four states, the adoption list has grown to forty-four American States; six Canadian Provinces; the British Territory of Bermuda; the Australian State of Queensland; the International Civil Aviation Organization Training Centre in Beirut, Lebanon; the Department of National Defence of Canada; the Department of the Army of the United States; the Department of the Navy of the United States; the United States Air Force; the United States Bureau of Indian Affairs; the United States General Services Administration, and the National Aeronautics and Space Administration (NASA). Representatives from the various adopting agencies serve as a voluntary group of individuals who govern policies, recommend procedures, and validate material before it is published. Most of the representatives are members of other international fire protection organizations and this meeting brings together men from several related and allied fields, such as:

> . . . key fire department executives and drillmasters,
> . . . educators from colleges and universities,
> . . . representatives from governmental agencies,
> . . . delegates of firefighter associations and organizations, and
> . . . engineers from the fire insurance industry.

This unique feature provides a close relationship between the International Fire Service Training Association and other fire protection agencies, which helps to correlate the efforts of all concerned.

The publications of the International Fire Service Training Association are compatible with the National Fire Protection Association's Standard 1001, "Fire Fighter Professional Qualifications (1974)," and the International Association of Fire Fighters/International Association of Fire Chiefs "National Apprenticeship and Training Standards for the Fire Fighter." The standards are an effort to attain professional status through progressive training. The NFPA and IAFF/IAFC Standards were prepared in cooperation with the Joint Council of National Fire Service Organizations of which IFSTA is a member.

In order to formally adopt IFSTA training material, the firefighters should first request it through their training agencies. While adopting does not obligate any sponsoring agency, it still profits by sending a representative to the Annual IFSTA Validation Conference. This representation permits a voice and a vote on the ever-changing policies, techniques, and procedures which must be reflected in each publication. Adoption further adds prestige to any training program and this, in turn, adds prestige to IFSTA publications.

The International Fire Service Training Association meets each July at Oklahoma State University, Stillwater, Oklahoma. Fire Protection Publications at Oklahoma State University publishes all IFSTA training manuals and texts. This department is responsible to the executive board of the association. While most of the IFSTA training manuals can be used for self-instruction, they are best suited to group work under a qualified instructor.

PREFACE

As technology moves forward fire protection devices are being developed and installed to immediately detect combustion and deliver a warning or initiate automatic suppression. Numerous fire extinguishing systems are now available that operate at the first sign of a fire, thus providing maximum protection and minimum losses.

Many systems have been in use for almost a century. However, new technology is ever improving the capability of these types of systems. Other systems using space-age developments are just off the drawing board.

The spiraling costs of construction, replacement, insurance and fire suppression coupled with the costs of business interruption have created a cost effective atmosphere for the installation of private fire protection and detection systems. The fire department in its roles of educator, consultant, inspector and user must be totally familiar with the many systems not only installed within its jurisdiction, but also available. As codes are amended, the need for automatic or manual fire protection and detection must be included or upgraded as necessary.

Recognizing that these devices are in greater use and that the fire service needed additional information and training in these areas, the Executive Board of the International Fire Service Training Association voted to publish a manual providing this data. The first edition of **Private Fire Protection and Detection** is a result of the Board's decision and numerous hours of research and compilation of information by editorial committees.

Private Fire Protection and Detection is not a highly technical manual of engineering principles, but presents the material in a practical, straightforward manner. The requirements of NFPA Standard No. 1001, *Fire Fighter Professional Qualifications,* for all levels of firefighters are met by this manual. Included are fire department operations and testing procedures for the various systems to increase the value of the manual to firefighters and inspectors.

The editorial staff would like to express its appreciation to the Continuing Committee who were appointed by the IFSTA Executive Board. These men made valuable suggestions and contributions and were instrumental in the final validation of the manual.

Chairman	**Vice Chairman**
Frank McGurn	Robert M. Porter
National Automatic Sprinkler and Fire Control Assn., Inc.	Bancroft Fire Dept.
Crystal Lake, Illinois	Lakewood, Colorado
Pat Brock	Richard Engle
Fire Protection and Safety Technology	International Harvester Co.
Oklahoma State University	Rock Island, Illinois

To the loyal and hard-working committee members who supported the Continuing Committee through numerous meetings and days of concentrated effort and who contributed to group discussion from their talents and knowledge, we also express our thanks.

Curtis K. Bell	Dale F. Janes	James Parman
Larry Borgelt	Eugene Kiefer	Kenneth L. Stanton
Jimmy D. Dickey	Laurence Lee	Edward D. Steiner
Richard W. Giles	William J. Matheny	James W. Townley
William L. Hanbury	Robert H. Noll	

Larry Borgelt and Pat Brock of the committee lent considerable experience, time and expertise to this manual during the critical production phases. Their additional efforts were extremely beneficial and deeply appreciated by the editorial staff.

Although not officially committee members, these individuals contributed greatly to the manual with their technical research and expertise.

Patrick Knight David Williams
Tim Perkins T.K. Gillmore
Robert Rieger

Our gratitude is also extended to the following individuals whose contributions made final publication of this manual possible.

Velinda Baker, E. T. Brown III, Pat Knight, Glenn Wagner and Don Davis for their assistance in obtaining the supporting photographs.

Nancy Wilson, Ann Moffat and Carol Smith for their assistance in phototypesetting, layout and proofreading of the manual.

The following organizations and associations generously granted permission to reprint copyrighted material, provided photographs for inclusion in the manual, or provided equipment in order for the IFSTA staff to take photographs used throughout the manual. To these organizations, we are grateful.

The Ansul Company
"Automatic" Sprinkler Corporation of America
Grinnell Fire Protection Systems Company, Inc.
National Fire Protection Association
Insurance Services Offices
The Reliable Automatic Sprinkler Co., Inc.
The Viking Corporation
School of Fire Protection and Safety Engineering Technology
Firefighter David Zink, Stillwater Fire Department
Dept. of Safety, Oklahoma State University
Chemtron Corporation Cardox Division
Kidde Belleville Division of Walter Kidde and Co., Inc.
Seaway Pipeline, Inc.

Note: All measurements are given in English units in the text. Appendix A is a treatise on metric units and contains the text conversions referenced by chapter and page.

CONTENTS

	INTRODUCTION	1
1	**AUTOMATIC SPRINKLER SYSTEMS**	5
	Valve of Sprinkler Systems for Life Safety	9
	Sprinkler Heads	10
	Sprinkler Temperature Ratings	15
	Control and Operating Valves	16
	Waterflow Alarms	19
	Water Supply	21
	The Wet-Pipe System	26
	Retarding Devices	27
	Operation of the System	29
	Using Sprinkler Stops	30
	Restoring the System	30
	Inspecting the System	31
	Wet-Pipe System Testing	35
	Frequency of Inspecting and Testing Wet-Pipe Sprinkler Systems	37
	Dry-Pipe Sprinkler System	37
	Accelerators and Exhausters	40
	Operation of the System	41
	Restoring the System	41
	Inspecting the System	42
	Dry-Pipe System Testing	44
	Frequency of Inspecting and Testing Dry-Pipe Systems	48
	Pre-Action System	48
	Deluge Sprinkler System	48
	Operation of the System	49
	Inspecting Deluge or Pre-Action Systems	52
	Frequency of Inspecting and Testing Deluge and Pre-Action Systems	55
	Sprinkler System Alarm Attachments	55
	Residential Sprinkler Systems	56
2	**SPECIAL EXTINGUISHING SYSTEMS**	59
	Carbon Dioxide Extinguishing Systems	60
	Fire Department Operations	67
	Inspecting and Testing of the System	67

Halogenated Agent Extinguishing Systems		68
System Description		70
Fire Department Operations with Halon		71
Inspecting the System		72
Dry- and Wet-Chemical Systems		73
Dry Chemicals		73
Dual or Combined Agents		77
Wet Chemicals		79
Inspecting and Testing Chemical Systems		80
Foam Systems		81
Chemical Foams		81
Protein Foams		82
Synthetic Foams		83
Foam Proportioners		85
Foam Generators		90
Fixed Foam Systems		94
Foam-Water Systems		98
High Expansion Foam Systems		100
Foam Applications		102
3	**STANDPIPE AND FIRE EXTINGUISHER SYSTEMS**	**103**
	Standpipe Systems	105
	Classes of Standpipe Systems	105
	Water Supply for Standpipe Systems	106
	Types of Standpipe Systems	107
	Pressure-Reducing Valves	107
	Standpipes in High-Rise Buildings	108
	Fire Department Connections	109
	Inspecting and Testing of Standpipes	110
	Fire Department Operations with Standpipes	111
	Outside Private Protection	114
	Fire Extinguishers	116
	Classification of Fires	116
	New Marking Systems	117
	Rating of Extinguishers	118
	Factors Influencing the Effectiveness of Fire Extinguishers	120
	Distribution of Extinguishers	121
	Inspecting Fire Extinguishers	123
4	**FIRE DETECTION AND ALARM SYSTEMS**	**125**
	Fire Detection and Alarm Systems	126
	Local Alarms	128

Proprietary Systems	134
Central Station	135
Remote Station	136
Municipal Fire Alarm Systems	137
Alarm Initiating Systems	139
Manually Activated Devices	139
Products of Combustion Detectors	140
Heat Detectors	141
Visible Products of Combustion Detectors	148
Invisible Products of Combustion Detectors	150
Light Detectors	152
Extinguishing-System Alarm-Initiating Devices	153
Telephone Dialers	153
Supervisory Circuits	154
Multiplex Systems	155
Inspecting and Testing of Alarm and Detection Devices	156
Inspecting	156
Manual Pull Stations	157
Detector Testing	158
Inspecting Control Equipment	159
APPENDIX A	**161**

Dedication

This manual is dedicated to the members of that unselfish organization of men and women who hold devotion to duty above personal risk, who count sincerity of service above personal comfort and convenience, who strive unceasingly to find better ways of protecting the lives, homes and property of their fellow citizens from the ravages of fire and other disasters . . . **The firefighters** *of All Nations.*

INTRODUCTION

Private fire protection includes various privately owned devices and equipment which have been installed or located within or about the property of the owner to deal with the outbreak of fire. These devices and equipment can be either manually or automatically operated. Private fire protection further includes employees who are trained to use these devices or equipment in the accomplishment of fire prevention or fire fighting duties. The purpose of private fire protection is to provide a means by which fires may be prevented or attacked in their incipient phase and/or controlled until public fire protection can arrive. There are many types of private fire protection systems and a short summary of each appears below.

Sprinkler systems were designed, in their basic form, over 100 years ago and prevalent fire losses made their perfection necessary. Today the automatic sprinkler system is an unsurpassed fire protection device. Annual records have revealed that in buildings where automatic sprinklers were installed 96 percent of all fires were controlled or extinguished by these systems. Of the remaining fires that were not controlled in sprinkler-equipped buildings, failure was due to improper maintenance, inadequate or shut off water supply, incorrect design, obstructions and partial protection.

Sprinklers — Over 100 years of unsurpassed protection

Although they cannot take the place of an automatic sprinkler system, standpipe systems are designed to provide a quick and convenient means for operating fire streams on all stories of buildings. Depending on the type installed, the standpipe system may be used by firefighters, by occupants, or both and should be ready for use at all times.

Special-agent fixed-extinguishing systems are used in those situations where water-application fixed systems, i.e. automatic sprinklers, are not applicable. In these instances, protection must still be provided safely and effectively. This can be accomplished by the use of carbon dioxide (CO_2), halogenated agents or a dry-chemical fixed system among others.

Carbon dioxide systems can be used to extinguish or control fires in either flammable liquids or gases and electrical equipment. Areas for their use would include flammable liquid dip tanks, engine test cells and computer installations. Since carbon dioxide is a gas when discharged, it will leave no residue, which means that cleanup time is minimal. Equipment is not damaged by the use of the agent; but because of the large carbon dioxide concentrations needed to extinguish fires in some instances, a health hazard does exist. For this reason a predischarge alarm is incorporated into some systems.

Halogenated-agent fixed systems can be used on fires in Class A, B or C materials. Like carbon dioxide it leaves no residue and is sometimes referred to as a "clean agent" system. Unlike carbon dioxide the concentrations needed for extinguishing fires under normal circumstances only run between three to seven percent. Although this allows people to be in the area of discharge of the halogenated agent for a short period of time with no ill effects, occupancy of the area during discharge is not recommended. Halogenated agents are highly recommended for areas of high value that are easily susceptible to damage such as computer rooms, record storage and fur vaults.

Dry-chemical fixed systems are primarily for use on flammable liquid fires and electrical equipment, especially where flammable liquids are contained in electrical equipment such as an oil-filled transformer. It is not recommended, however, for delicate electronic equipment because of the cleanup problem involved.

The hand hoseline systems contain a large tank for storing the dry chemical agent, a cylinder or "bank" of cylinders filled with inert gas for expelling the agent, hose and special nozzles. Handline systems range in size from 150 pounds to 2,000 pounds.

All three of the above systems can be designed for a total flooding application as well as local application. The type of system used, as well as its arrangement, will depend on such factors as the hazard, its size, organization and economics.

Special-agent systems are designed for local and total flooding applications

Foam-water extinguishing systems provide basically the same general protection as the standard deluge-type automatic sprinkler system. The foam-water system does, however, have a decided advantage where the hazard involves flammable liquids or adequate floor drainage cannot be provided. One advantage is the foam's fast smothering action. Where drains are not available or water-susceptible stock is involved, the low water content of this foam solution will reduce absorption and lessen the weight load on floors. The system is most commonly used in facilities which store and process flammable liquids, such as chemical plants, solvent extraction plants, distillation plants and refineries. The major advantages of this system are the characteristics of the foam agent to suppress flammable liquid fires by blanketing, to flow better and to control possible ignition by covering the surface of liquid spills.

The most common of all private fire protection is the portable fire extinguisher. Fire prevention and suppression personnel and maintenance personnel must have an intimate knowledge of the characteristics and applicability of each type, as well as the classification and distribution requirements as explained in NFPA Standard No. 10, *Portable Fire Extinguishers,* and IFSTA 101, **Forcible Entry, Rope and Portable Extinguisher Practices.** A knowledge of the application of NFPA Standard No. 10-L, *Model Enabling Act, Portable Fire Extinguishers*, is also important if such legislation has been enacted.

Fire detection and alarm systems are used to provide signals to alert building occupants and/or organized fire protection units, and to operate fire protection system components. All detection systems use some type of device which is sensitive to heat, flame or products of combustion. Automatic fire detection and alarm systems with supplemental manual fire alarm pull-stations should be installed in buildings for protection of life or in isolated and/or high life-hazard facilities where automatic sprinkler protection would normally be provided but is not economically or technically feasible. With more emphasis being placed on early detection and regulations being enacted that require detectors to be installed in dwellings, it becomes increasingly necessary for all fire service personnel to become well-versed in these devices. **NOTE:** Automatic fire detection systems are not acceptable substitutes for automatic sprinkler systems.

Increased emphasis on early detection

A vast number of variations in types, makes and models of fire protection and detection equipment are recognized. Systems, equipment and procedures described herein have been used to illustrate typical installations. Generally, the more complicated or sophisticated a system is, the more important skillful maintenance becomes. It is recommended that the manufacturer's technical data be consulted for specific answers to questions concerning design, installation, operation and maintenance of components or systems.

Review manufacturer's technical data for specifics

The installation, maintenance and testing of fire protection and detection equipment must be in compliance with local, state and/or federal codes, ordinances and standards. Many installations also follow applicable NFPA, UL or FM guidelines. Materials contained in the manual are intended to be descriptive in nature and should not be used as the authority for examination of any legally constituted requirement.

Fire service personnel will find that materials in this manual will help to bring about a better understanding of private fire protection and detection system installations. Those personnel involved with fire protection management, safety and maintenance should also be better able to understand the concern of the public fire services for proper maintenance of this equipment. Often, building codes and ordinances are relaxed due to the installation of fire protection and detection systems. This could constitute a serious threat to the safety of the public and fire service personnel in the event a fire occurs.

AUTOMATIC SPRINKLER SYSTEMS

- Sprinkler Heads
- Wet-Pipe Systems
- Dry-Pipe Systems
- Pre-Action and Deluge Systems
- Inspecting and Testing

NFPA STANDARD 1001

Firefighter I

3−13 Sprinklers.

3−13.1 The firefighter shall identify the fire department sprinkler connection and water motor alarm.

3−13.2 The firefighter shall connect hoseline(s) to a fire department connection of an automatic sprinkler system.

3−13.3 The firefighter, when given a sprinkler head in serviceable use, shall demonstrate knowledge of how the automatic sprinkler head opens and releases water.

3−13.4 The firefighter, when given the necessary equipment, shall effect a temporary stop of the flow of water from a sprinkler head.

Firefighter II

4−14 Sprinklers.

4−14.1 The firefighter shall identify the MAIN DRAIN valve on an automatic sprinkler system.

4−14.2 The firefighter shall open and close a MAIN DRAIN valve on an automatic sprinkler system.

4−14.3 The firefighter shall identify the MAIN CONTROL valve on an automatic sprinkler system.

4−14.4 The firefighter shall operate a MAIN CONTROL valve on an automatic sprinkler system from "open" to "closed" and then back to "open."

4−14.5 The firefighter shall demonstrate knowledge of the value of automatic sprinklers in providing safety to life of occupants in a structure.

4−14.6 The firefighter shall identify and explain the dangers of premature closure of a sprinkler MAIN CONTROL valve, and of using fire hydrants to supply fire hose streams when the same water system is supplying the automatic sprinkler system.

4−14.7 The firefighter shall identify the difference between an automatic sprinkler system that affords complete coverage and a partial sprinkler system.

4−14.8 The firefighter shall identify at least three sources of water for supply to an automatic sprinkler system.

4−14.9 The firefighter shall identify the following:

    (a) Wet sprinkler system

    (b) Dry sprinkler system

    (c) Deluge sprinkler system

4–14.10 The firefighter, when given the tools and sprinkler head, shall properly remove one head from the system and replace it with a head of the same type.

Firefighter III

5–9 Sprinklers.

5–9.1 The firefighter, given an alarm valve of an automatic sprinkler system, shall demonstrate the operation of the valve.

5–9.2 The firefighter, given twelve various sprinkler heads, shall identify all of them correctly as to:

    (a) Temperature rating

    (b) Pendant or upright

    (c) Special types

5–9.3 The firefighter shall identify the ALARM TEST valve on an automatic sprinkler system.

5–9.4 The firefighter, given an automatic sprinkler system, shall operate the ALARM TEST valve in such a manner as to actually test the system.

5–9.5 The firefighter, given a velocity drain valve or ball drip valve on the fire department connection of an automatic sprinkler system, shall demonstrate that the valve is operating and the pipe drained.

5–9.6 The firefighter, given a check valve on the fire department connection to an automatic sprinkler system, shall demonstrate the direction of flow of water through the valve.

5–9.7 The firefighter shall read and record the indicated pressures on all gauges provided on a standard WET automatic sprinkler system and name each gauge.

5–9.8 The firefighter shall read and record the indicated pressures on all gauges provided on a standard DRY pipe automatic sprinkler system and name each gauge.

5–9.9 The firefighter shall identify and explain the reliability of automatic sprinkler systems, and shall identify eight reasons for unsatisfactory performance.

5–9.10 The firefighter, by inspection of an automatic sprinkler system in a building, shall identify and explain if obstructions to sprinkler heads are present and what is the required clearance for the sprinkler head from obstructions.

Reprinted by permission from NFPA Standard No. 1001, *Standard for Fire Fighter Professional Qualifications.* Copyright © 1974, National Fire Protection Association, Boston, MA.

AUTOMATIC SPRINKLER SYSTEMS

Early types of sprinkler systems were rather crude and unreliable, but present-day systems have been perfected to the point that they are extremely reliable when properly maintained. The reduction of insurance rates for property that is equipped with sprinkler protection has been a very influencial factor in the growth of the number of installations in properties.

Most reliable of all fire protection devices

The automatic sprinkler heads and all component parts of the systems should be listed by a nationally recognized testing laboratory, such as Underwriters Laboratories or Factory Mutual. Automatic sprinkler systems are now recognized as the most reliable of all fire protection devices and an understanding of the system of pipes and valves and their operation is essential to the firefighter.

Automatic sprinkler protection consists of a series of devices so arranged that the system will automatically distribute sufficient quantities of water to either extinguish a fire or to hold it in check until firefighters arrive. Water is supplied to the sprinkler heads through a system of piping. The sprinkler heads can either extend from exposed pipes or protrude through the ceiling or walls from hidden pipes.

The process of spacing sprinkler heads in a building must conform to well-established and tested standards, such as NFPA Standard No. 13, *Installation of Sprinkler Systems*. Standards are also set for the size of pipe to be used, the proper method of hanging the pipe and all other details concerning the installation of a sprinkler system. The design of automatic sprinkler systems is based upon the assumption that only a portion of the sprinkler heads will be opened during a fire. Most ordinary public waterworks systems could not be expected to adequately supply 500 or 1,000 operating sprinklers. This statement emphasizes the fact that a sprinkler system must be properly designed and an adequate source of water must be provided. Water volume and pressure must be adequate for the number of sprinkler heads that are calculated to operate at one time in any given building.

Sprinkler systems rarely fail to operate

In general, reports reveal that only in rare instances do automatic sprinkler systems fail to operate. When failures are reported, the reason for failure is usually due to a lack of water or from the water supply being shut off instead of failure of the actual sprinklers. It has been established that approximately $150 billion worth of property is protected by automatic sprinklers in the United States. Records and reports further reveal that the annual fire loss in this protected property would be about 90 percent higher if the property was not protected by sprinklers.

On an average, about 96 percent of all fires in sprinklered buildings are either extinguished by sprinklers or held in check until they are completely extinguished by firefighters. Thirty-five percent of the remaining four percent of fires that were not controlled by the sprinkler system was due to closed valves, or closing control valves before complete extinguishment. The remaining 65 percent of the failures was due to various reasons such as poor or improper maintenance, improper design, hazards of the occupancy, distribution obstructions, deficient water supplies or only partial sprinkler protection. Another interesting feature of sprinklered buildings is that 35 percent of the fires that were controlled by sprinkler operations were controlled by the operation of only one sprinkler head.

VALUE OF SPRINKLER SYSTEMS FOR LIFE SAFETY

The life safety of building occupants is enhanced by the presence of a sprinkler system because it discharges water directly on the fire while it is relatively small. Since the fire is extinguished or controlled in an early stage, the combustion products are limited. Most fire fatalities in a sprinklered building are caused by asphyxiation either because of a small fire which does not generate sufficient heat to fuse a sprinkler, or because the victim had suffered fatal injuries by the time the sprinkler operated. Other fatal instances include sleeping, intoxicated, or handicapped persons and ignition of clothing or bedding which causes fatal burns; however, the sprinkler system protected the lives of persons in other parts of the building.

Automatic sprinklers enhance life safety

An individual's life safety in a sprinklered high-rise building is increased many times. The probability of occupant rescue in an unsprinklered building is lower because of manual fire suppression.

In addition to assisting in the planning of installed fire protection systems and devices, the fire department may witness initial inspection and testing of those systems which are supervised by the installer. Routine checks and maintenance are often performed by plant personnel, such as checking for closed valves and performing periodic tests. An effective arrangement is for the occupant to establish a continuing contractual agreement with the company that installed the system to inspect the system, accomplish periodic tests and perform required maintenance. It is the responsibility of the owner to maintain the system in efficient service. If the systems are maintained by contract or occupant personnel, the fire department should assist by making spot checks to insure that the system is being properly maintained. This should include witnessing drain and alarm tests. If requested, the fire department should assist with the periodic review of any existing maintenance contracts.

A sprinkler system piping layout, as shown in Figure 1-1, consists of different size pipe. The system starts with a feeder main into the sprinkler valve. A riser is a vertical pipe supplying the sprinkler system, generally a one-way check valve is installed. System piping decreases in size from the riser outward. The pipes connecting the riser to the cross-mains are known as the feed mains. The cross-main directly services a number of branch lines on which the sprinklers are installed. Cross-mains extend past the last branch lines and are capped to facilitate flushing. The entire system is supported by hangers and clamps. All pipes in dry systems are pitched to help drain the system back toward the main drain.

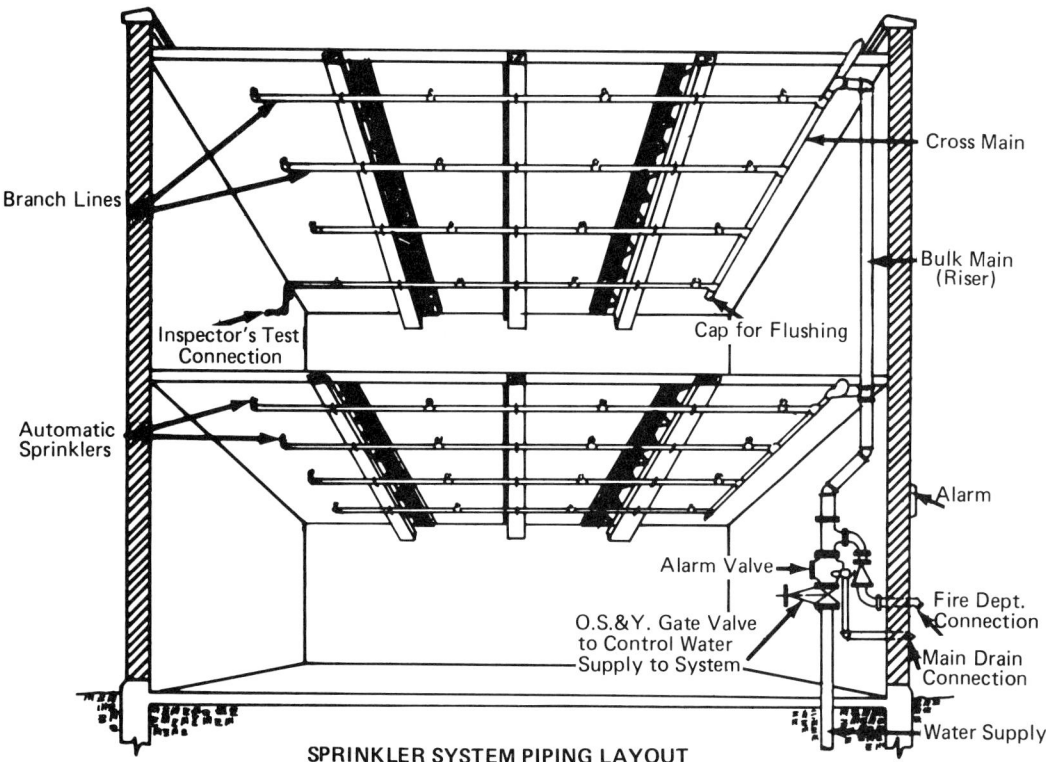

Figure 1-1 Typical piping schematic of an automatic sprinkler system, showing important parts.

SPRINKLER HEADS

Automatic sprinklers, often called "sprinkler heads" or just "heads," discharge water after the release of a cap or plug which is activated by some heat-responsive element. This head may be thought of as a fixed-spray nozzle that is operated individually by a thermal detector. There are numerous types and designs of sprinkler heads. Only the more common types will be covered in this chapter.

Automatic Sprinkler Systems

 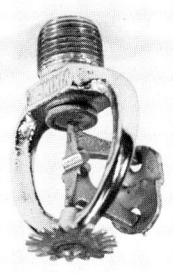  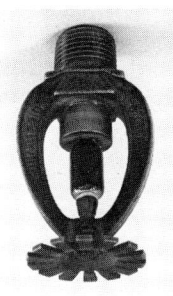  

Figure 1-2 Upright and pendant heads with various types of release mechanisms.

Installed heads in wet- and dry-pipe systems are kept in a closed position by various devices (Figure 1-2). Three of the most commonly used release mechanisms are fusible links, glass bulbs and chemical pellets, all of which fuse or open in response to heat. The major parts of the fusible-link sprinkler head are shown in Figure 1-3. As illustrated in Figure 1-4 the solder melts in the fusible link at a predetermined temperature; then the lever arms are released and spring clear of the head. As the lever arms pop out, the seated valve cap is released, which permits the water to flow. The standard orifice over which the cap is held is one-half inch in diameter. A deflector is attached to the sprinkler frame. Discharge water is directed against this to convert it into a spray. The fusible-link sprinkler head in operation is shown in Figure 1-5.

FUSIBLE LINK SPRINKLER COMPONENTS

Figure 1-3 Sprinkler heads are made up of many parts attached to the frame. The individual parts keep the head closed at normal temperatures, but fall away when the temperature exceeds the rated temperature of the head.

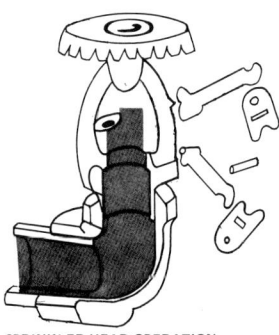

Figure 1-4 An upright sprinkler head illustrating the activation of the fusible link.

SPRINKLER HEAD OPERATION

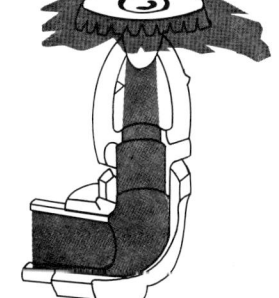

Figure 1-5 Illustration of how water flows through the head after the fusible link has actuated.

SPRINKLER HEAD OPERATION

OCCUPANCY HAZARD	ORIFICE	PROTECTION AREA	
Light	½-inch	130 to 168 sq. ft. 200 or more sq. ft. permitted under some circumstances	Consider construction feature
	⅜-inch	65 to 100 sq. ft.	
Ordinary	½-inch	130 sq. ft. 15 ft. between branch lines and sprinklers	
		Except 100 sq. ft. for high piled stock	
Extra	½- or 17/32-inch	90 sq. ft. 12 ft. between branch lines and sprinklers	Depending on construction features and hazardous steps or processes

12 PRIVATE FIRE PROTECTION

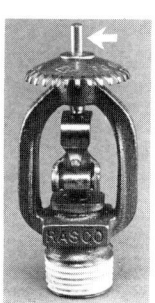 

Standard sprinkler heads are designed to discharge a spray of water downward in a hemispherical pattern. The upright-type standard sprinkler heads cannot be inverted for use in the hanging or pendant position because in the inverted position the spray of water would be directed toward the ceiling due to the design of the deflector. Pendant-type sprinklers should be used in locations where it is impractical to use sprinklers in an upright position. Some specialized sprinkler heads are shown and briefly explained in Figure 1-6.

Figures 1-6-1 & 1-6-2 The pintle (see arrow) on a sprinkler head signifies that the head has a ½-inch pipe thread connection; however, a differing orifice size.

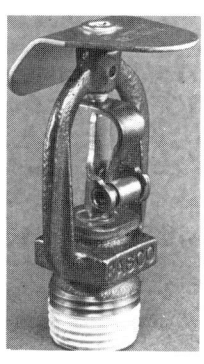

Figure 1-6-3 The vertical sidewall head is mounted close to the wall and the specially designed deflector directs a water spray away from the wall. This type of head is used to protect small rooms, hallways, atriums and vertical openings.

Figure 1-6-4 The extended coverage sidewall head is used when a long-reaching spray pattern is needed.

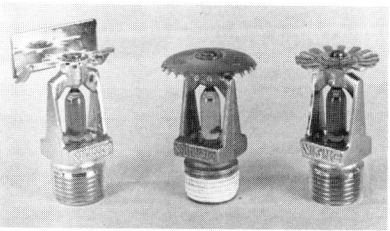

Figure 1-6-5 Decor heads are used in areas where appearance is a concern. They are available in several finishes and styles.

Figure 1-6-6 A variety of different size sprinkler heads are available.

Figure 1-6-7 A directional head is used in special areas for spray direction.

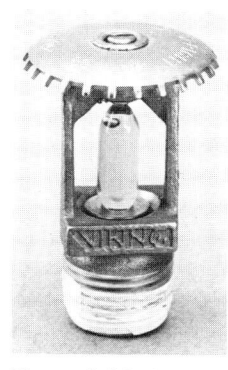

Figure 1-6-8 An intermediate-temperature decorative head with no marking on the frame.

Figure 1-6-9 Although not required, this decorative head has a blue dot in the center of the deflector to indicate the temperature rating.

Figure 1-6-10 This pendant head is mounted with an escutcheon plate to cover exposed piping.

Figure 1-6-11 The quick-response type of head is used where reaction time is critical. Heat collects in the bell-shaped heat collector and activates the head in a shorter time.

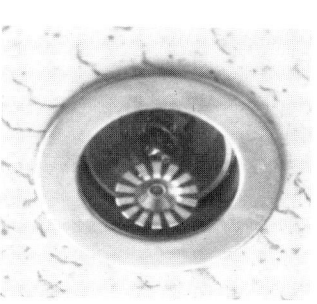

Figure 1-6-12 A recessed sprinkler is mounted flush with the deflector exposed. This type of head is used for its esthetic quality.

Figures 1-6-13 & 1-6-14 This flush-mounted wet pendant sprinkler is mounted where appearance is of concern. The activated head shows the extended position of the spray deflector.

Automatic Sprinkler Systems 13

Figures 1-6-15 & 1-6-16 These heads are designed for installation inside storage racks. The large deflector or shield prevents water discharging from overhead sprinklers from cooling the lower heads and delaying activation.

Figure 1-6-17 Coated sprinklers are used where the atmosphere could deter the operation of the head.

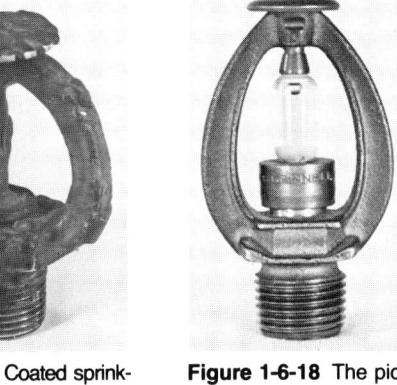

Figure 1-6-18 The picker head deflector will minimize the accumulation of foreign material.

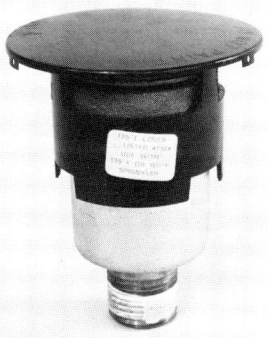

Figure 1-6-19 The unspoiler head is mounted flush with the ceiling for esthetic purposes. The cover drops when the temperature rises to approximately 135°F (57.2°C). The head fuses at a temperature of approximately 165°F (73°C).

Figure 1-6-20 Quick response actuator to attach to heads using utectic solder in the fusible link. The actuator sends hot gases directly on the solder link reducing the activation time.

Figure 1-6-21 Pilot heads in special hazards areas have their on-off action caused by a pressure variation in the pilot line actuator controlled by a detection system. This controls all heads in a deluge-type action.

Figure 1-6-22 An automatic shutoff-type head is used where there is concern about excessive water damage. The water is turned on and off automatically.

Figure 1-6-23 Open-nozzle heads are used in high hazard areas where a large amount of water is needed quickly. These four pendant heads have different-type deflectors for different directional sprays.

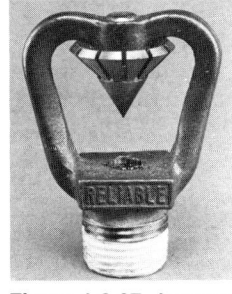

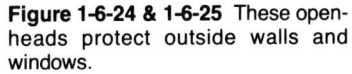

Figure 1-6-24 & 1-6-25 These open-heads protect outside walls and windows.

Figure 1-6-26 This chemical pellet on-off head after activation opens and closes by the expansion contraction of paraffin wax. Replace after use.

Figure 1-6-27 An open deluge head for special applications. Orifice size and deflector angle can be changed based upon need.

Figure 1-6-28 A cornice head is installed under cornices or roof ridges to create a narrow fan-shaped water pattern.

Standard sprinkler heads may be substituted for old-style heads in an existing system when it becomes necessary to change a head or to upgrade the protection provided. Old-style heads should not be used to replace standard heads because the old-style heads are equipped with deflectors which direct a large percentage of the discharge to a circular area of the ceiling above the sprinkler with the remainder of the discharge being directed downward in a roughly conical pattern (Figure 1-7). The primary objection to the old-style head is that the water falls in large drops from the ceiling. The new-style heads provide for better distribution of the water in the form of a dense fog or fine spray. Sprinkler heads may have a cage installed for protection against mechanical injury. all sprinkler heads are stamped, indicating the temperature rating, date of manufacture, name of manufacturer and position head is to be installed.

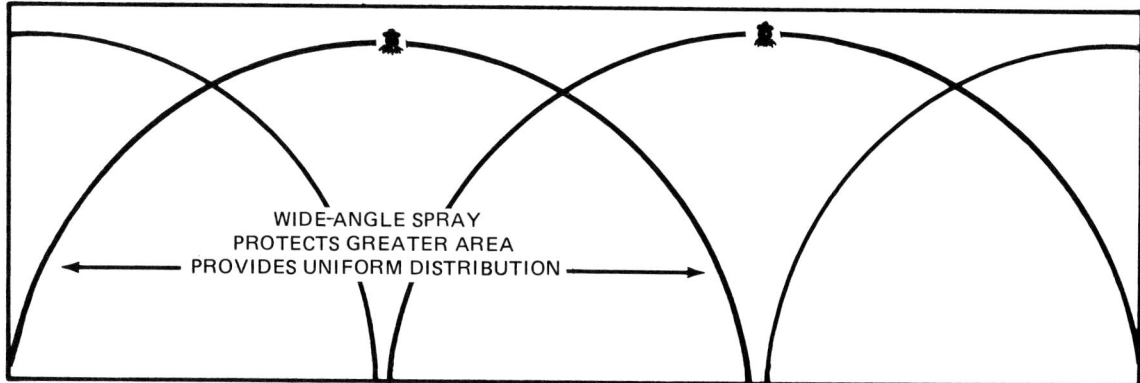

DISCHARGE PATTERN OF STANDARD SPRINKLERS

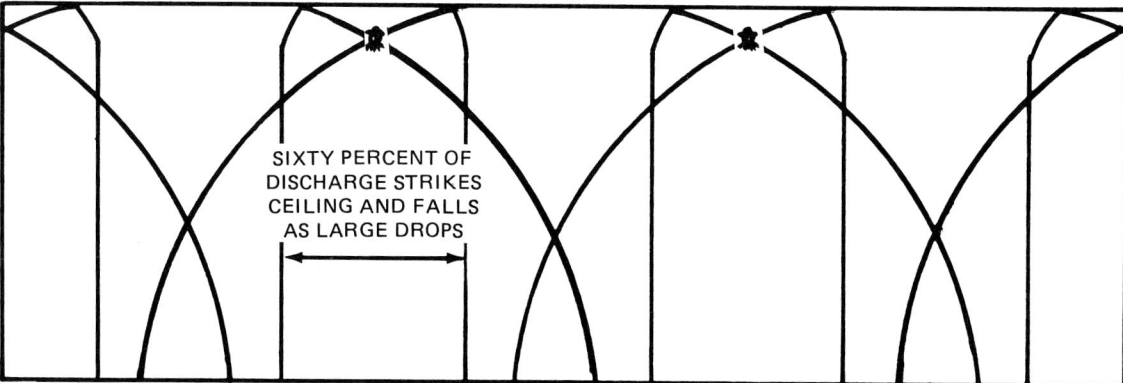

DISCHARGE PATTERN OF OLD-TYPE SPRINKLERS

Figure 1-7 Compare the discharge patterns of the old- and new-style sprinkler heads. Notice the uniform discharge pattern and coverage of the new style.

Special-type sprinkler heads with corrosive-resistant coatings are available and should be installed in areas where chemical, moisture or other corrosive vapors exist. Without this special coating, the operational parts of the head will corrode and may become inoperative in a very short time (Figure 1-6).

A storage cabinet should be installed in the area protected by the sprinkler system to house extra heads and a sprinkler head wrench. Normally these cabinets hold a minimum of six heads for small systems. It is necessary to use a sprinkler wrench and to be careful when changing heads to prevent damage (Figures 1-8 and 1-9).

Figure 1-8 A typical storage cabinet for storage of spare sprinkler heads.

Figure 1-9 Assorted sprinkler head wrenches.

SPRINKLER TEMPERATURE RATINGS

The sprinkler head used for a given application should be based on the maximum temperature expected at the level of the sprinkler under normal conditions, and the anticipated rate of heat release that would be produced by a fire in the particular area. The temperature rating shall be indicated by color coding the frame arms of the head except for coated sprinklers and decorative heads (Chart 2). Coated sprinklers have colored frame

TEMPERATURE RATINGS °F.	TEMPERATURE CLASSIFICATION	FRAME COLOR
135 to 170	Ordinary	Unpainted (or partly black or chrome)*
175 to 225	Intermediate	White*
250 to 300	High	Blue
325 to 375	Extra High	Red
400 to 475	Very Extra High	Green
500 to 575	Ultra High	Orange

* The 135° sprinklers of some manufacturers are half black and half painted. The 175° sprinklers of these same manufacturers are yellow.

16 PRIVATE FIRE PROTECTION

Figure 1-10 An indicating control valve that is electronically supervised is shown in the open position.

arms, coating material, or use a colored dot on the top of the deflector. Decorative sprinklers such as plated or ceiling sprinklers are not required to be color coded; however, some manufacturers use a dot on the top of the deflector (Figure 1-6).

CONTROL AND OPERATING VALVES

Every sprinkler system is equipped with a main water control valve and various test and drain valves. Control valves are used to cut off the water supply to the system when heads must be replaced, when maintenance is performed or when operation must be interrupted. The main control valve should always be returned to the open position after maintenance is completed. These valves are located between the source of water supply and the sprinkler system (Figure 1-10). These control valves are an "indicating" type and are manually operated. An indicating control valve is one that shows at a glance whether it is open or closed. These valves are usually located immediately under the sprinkler alarm valve, the dry-pipe or deluge valve, or outside of the building near the sprinkler system that it controls. Separate control valves are required for each system. Each valve should be chained open or have a closed-valve alarm.

Water supply control valves for sprinkler systems may be indicating gate valves or indicating butterfly valves. The mechanism consists of a close-tolerance gate that slides or turns across the waterway in the water main. There are three common types of indicator control valves used in sprinkler systems. One of these valves is an outside screw and yoke valve, usually called an OS&Y valve (Figure 1-11). This valve has a yoke on the outside with a threaded stem which controls the opening and closing of the gate. The threaded portion of the stem is out of the yoke when the valve is open and inside the yoke when the valve is closed.

Other types of control valves used are the post indicator valve (PIV), the wall post indicator valve (WPIV) and the post indicator valve assembly (PIVA). The PIV is a hollow metal post that is attached to the valve housing. The valve stem is inside of this post; on the stem is a movable target on which the words "OPEN" and "SHUT" are printed. The operating handle is fastened and normally locked to the post. When the valve is closed, the word "SHUT" appears at the opening. A PIV with the operating handle in both the stored and locked position is shown in Figure 1-12. WPIV is similar to a PIV except that it extends through the wall with the target and valve operating nut on the outside of the building (Figure 1-13). PIVA is similar to the PIV except that the valve used is a butterfly type while the PIV and the WPIV use a gate valve (Figure 1-14).

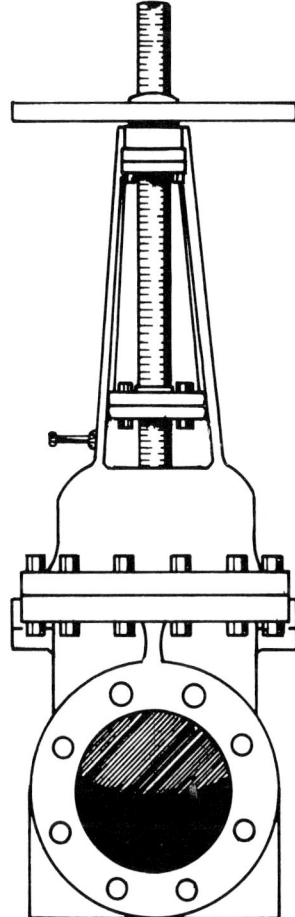

Figure 1-11 An illustration showing an OS&Y valve that is practically open.

Figure 1-13 A wall post indicator valve locked in the open position.

Figure 1-12 A post indicator valve in the open position. This type of indicating valve will display the word open or shut depending on the valve position.

Figure 1-14 A post indicator valve assembly with the butterfly type of valve.

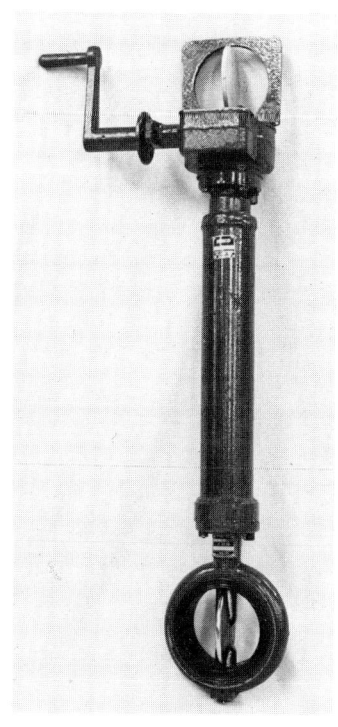

18 PRIVATE FIRE PROTECTION

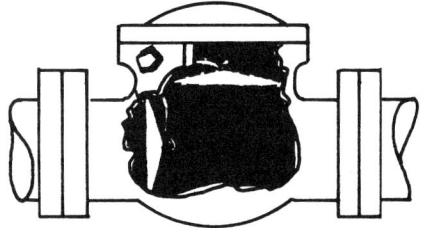

Figure 1-15 Two typical valves are shown here. Note the arrow on the casting that denotes the direction of the flow on the top valve and that the pivot casting would indicate the direction on the bottom valve.

In addition to the main water control valves, sprinkler systems will employ various operating valves such as globe valves, stop or cock valves, check valves and automatic drain valves. Globe valves and stop or cock valves are manually operated nonindicating valves. The globe valves are used for draining purposes and for test valves. Stop or cock valves are used for draining and for alarm silencing. Check valves are used to limit the waterflow to one direction. Check valves usually have an arrow printed on them to indicate the direction of water flow. If no arrow or indication of flow direction is identifiable on the exterior of the valve, a pivot casting should be on the end toward the fire department connection (Figure 1-15). If the arrow is pointing toward the fire department connection or the pivot casting is on the sprinkler riser side, the check valve is improperly installed. Water pouring from a fire department connection indicates the check valve is not seated properly. Improper seating of the valve can be caused by a bad rubber seat, obstructions or a sticking clapper. In any case, maintenance or replacement of the valve is indicated. Automatic drain valves serve to automatically drain piping when pressure is relieved from the valve. The alarm test valve is located on a pipe which connects the supply side of the alarm check valve to the retard chamber (Figure 1-16). This valve is provided to simulate actuation of the system by allowing water to flow into the retard chamber and operate the waterflow alarm devices.

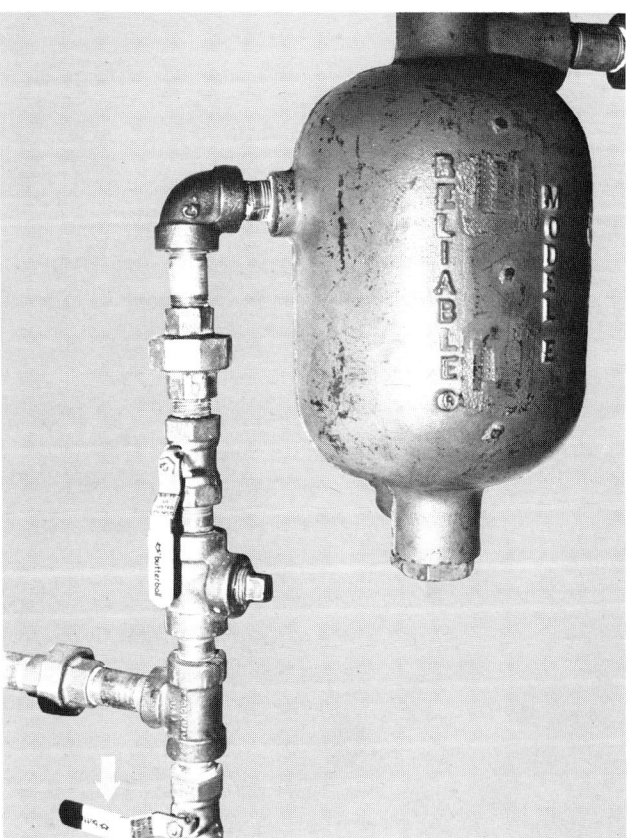

Figure 1-16 The alarm test valve is used to test the alarm devices by allowing water to them through the retard chamber without tripping the system valve.

Figure 1-17 The outside alarm portion of a water-motor gong.

Figure 1-18 This waterflow alarm transmits an electronic signal giving notification that water is flowing.

WATERFLOW ALARMS

Activation of fire alarms is accomplished by the operation of the alarm check valve, dry-pipe valve or deluge valve. These three valves will be discussed later. Sprinkler waterflow alarms are normally operated either hydraulically or electrically to warn of a waterflow. The hydraulic alarm is a local alarm used to alert the personnel in a sprinklered building or a passer-by that water is flowing in the system (Figure 1-17). This type of alarm uses the water movement in the system to branch off to a water motor, which drives a local alarm gong. The electric waterflow alarm is also employed to alert building occupants and, in addition, it can be arranged to notify the fire department. With this type of alarm, the water movement presses against a diaphragm which in turn causes a switch to operate the alarm (Figure 1-18). Immediate notification of the fire department is very important because support of the sprinklers is necessary although the system may control the fire. Prompt response by the fire department to support the sprinkler system will hasten extinguishment and reduce damage. In all cases, prompt notification is essential.

20 PRIVATE FIRE PROTECTION

Figure 1-19 This sectional riser has an electronically supervised OS&Y valve and an electric waterflow indicator.

Figure 1-20 The flow of water activates the vane in an electronic waterflow indicator. *(Courtesy of Insurance Services Office).*

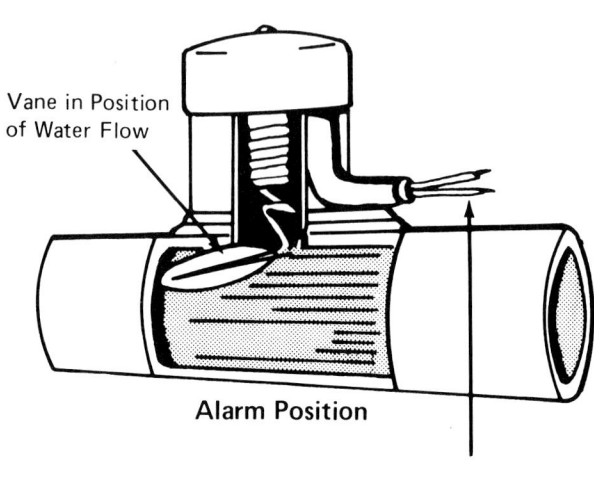

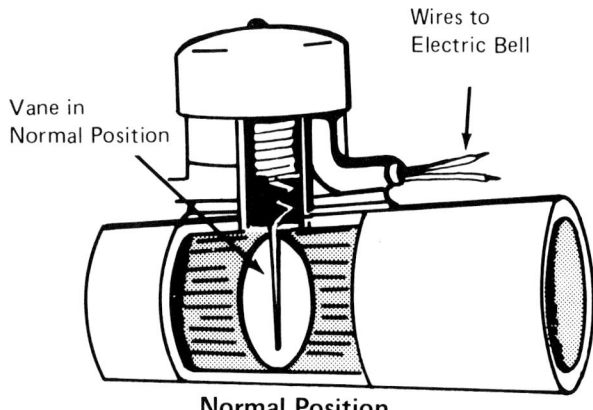

WATER FLOW DETECTOR

Waterflow indicators of the vane type are generally used in wet-system risers where there are no alarm-type check valves. These detectors are also used in floor laterals for localized notification of the flow of water in the sprinkler system. This arrangement may be found in multiple-story buildings. They serve to localize, by signal, waterflow in certain areas. In addition, sectional OS&Y valves can be installed to isolate sections of the sprinklered areas without interrupting the entire sprinkler system covered by one riser. An example of the arrangement of an area waterflow detector and sectional OS&Y valve is illustrated in Figure 1-19.

The vane-type indicator consists of a flexible plastic vane inserted through a hole drilled in the sprinkler riser or lateral, as shown in Figure 1-20. The extending end of the vane is connected to a switch mechanism encased in a housing mounted on the outside of the pipe. When water flows in the pipe, the vane is deflected, which operates the switch contacts and actuates the alarm transmitter or annunciator.

WATER SUPPLY

Every sprinkler system should have an automatic water supply of adequate volume, pressure and reliability. In some instances, a second independent water supply is not only desirable but required. A minimum water supply must be able to deliver the required volume of water to the highest sprinkler head in a building at a residual pressure of 15 psi. The minimum flow is established by the hazard to be protected, and is dependent upon the occupancy and fire loading conditions. A connection to a public water system that has adequate volume, pressure and reliability is a good source of water for automatic sprinklers. This type of connection is often the only water supply. A gravity tank of the proper size also makes a reliable primary water supply. In order to give the minimum required pressure, the bottom of the tank should be at least 35 feet above the highest sprinkler head in the building. Pressure tanks are another source of water supply and they are usually used in connection with a secondary supply. Pressure tanks are normally located on the top floor or on the roof of buildings. This type of tank is filled two-thirds full with water and it carries an air pressure of at least 75 psi. Adequate fire pumps that take suction from large reservoirs are used as a secondary source of water supply. When properly powered and supervised, these pumps may be used as a primary source of water supply.

Figure 1-21A A typical building-mounted fire department connection.

Figure 1-21B A fire department connection located away from the building.

One or more fire department connections through which the fire department can pump water into the sprinkler system may serve as an auxiliary water supply source (Figure 1-21). Under most conditions, especially when a large number of sprinklers are open or other water sources are weak, it is very important for firefighters to connect pumpers to these connections in order to boost the water volume and pressure. On a single-riser system, the fire department connection is attached on the sprinkler-system side of the main water supply valve. On multiple-riser systems, the fire department connection enters the supply piping between the main supply shutoff and the individual riser shutoffs. This design is so that in case the main water supply valve is inadvertently or otherwise closed, water can still be pumped into the system. On multiple systems, however, if sectional or floor valves are closed, the support of the fire department connection will be negated in that area. Fire department support through these connections is important enough to require the first or second pumper responding to hook up to these connections.

In addition, a check valve is installed in the feeding water main to prevent fire department pumpers from pumping back into the supply main instead of into the system. Clapper valves in the fire department connection prevent water from flowing out of the fire department connection when only one hoseline is connected. The check valve and ball drip in the fire department connection line are installed to keep the fire department connection dry to prevent freezing. Figures 1-22, 1-23 and 1-24 show the proper arrangement of the fire department connection to the sprinkler system.

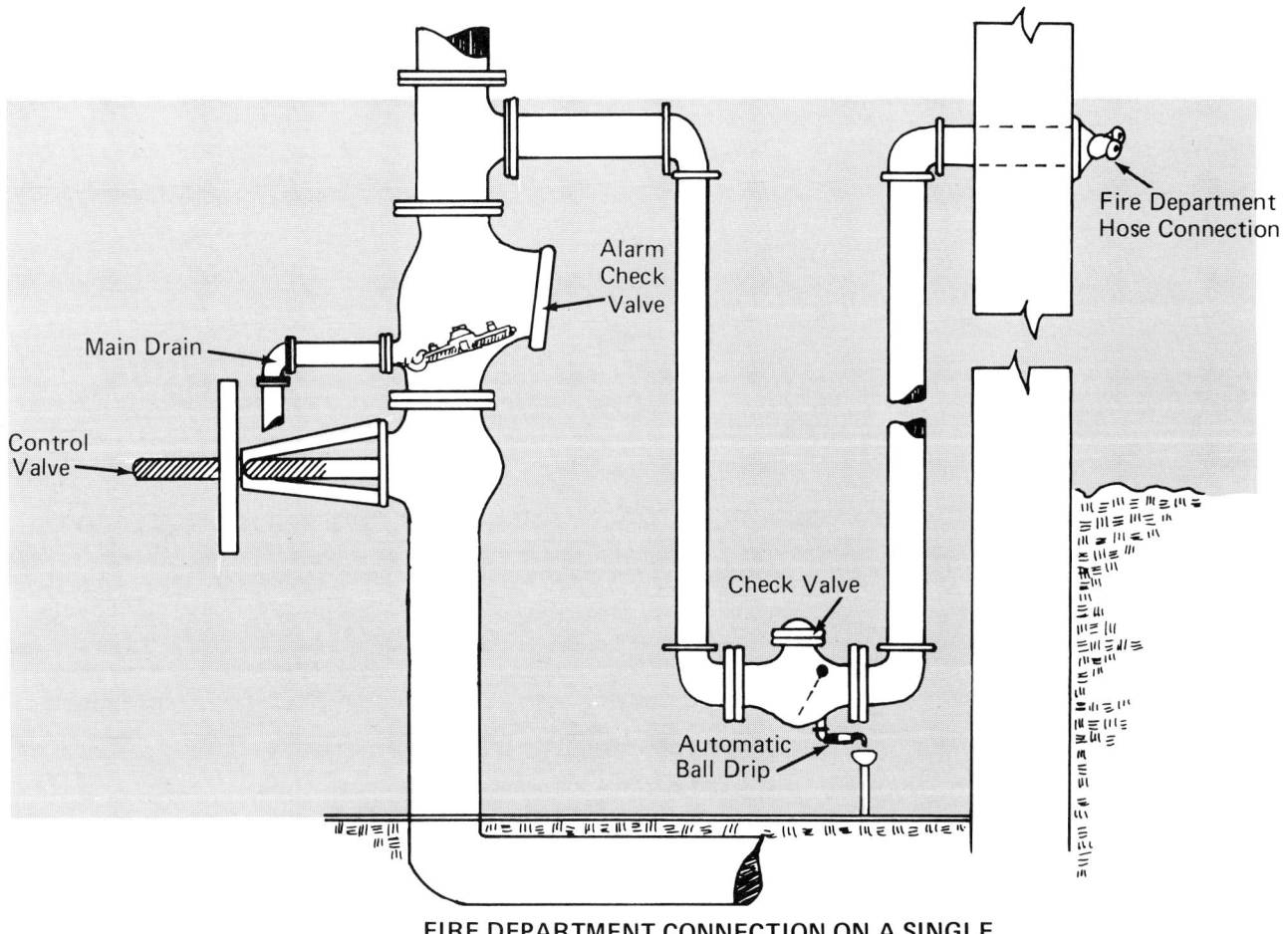

FIRE DEPARTMENT CONNECTION ON A SINGLE RISER WET PIPE SYSTEM

Figure 1-22 This piping schematic of a wet-pipe system illustrates that the fire department connection to the main riser is directly above the alarm check valve.

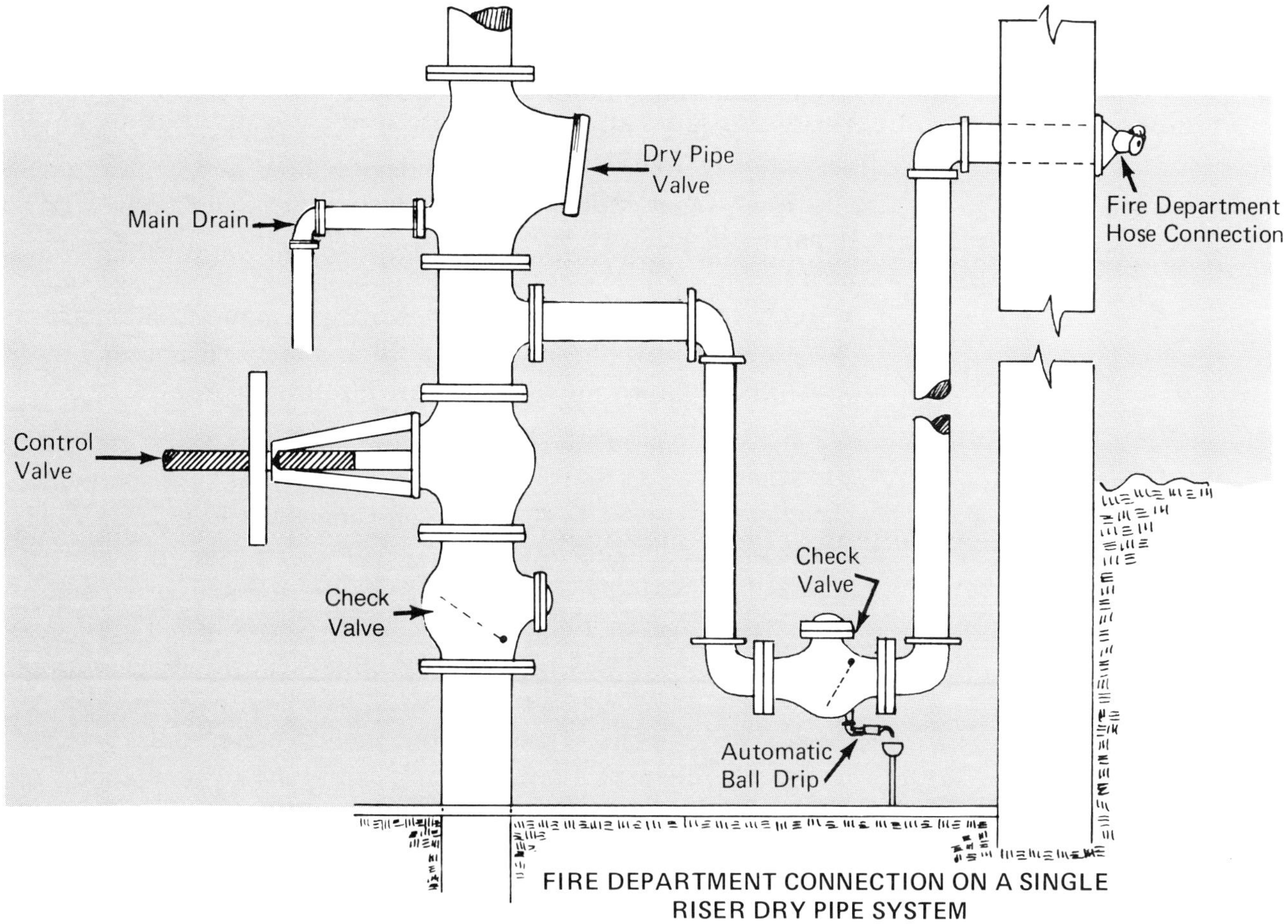

Automatic Sprinkler Systems **23**

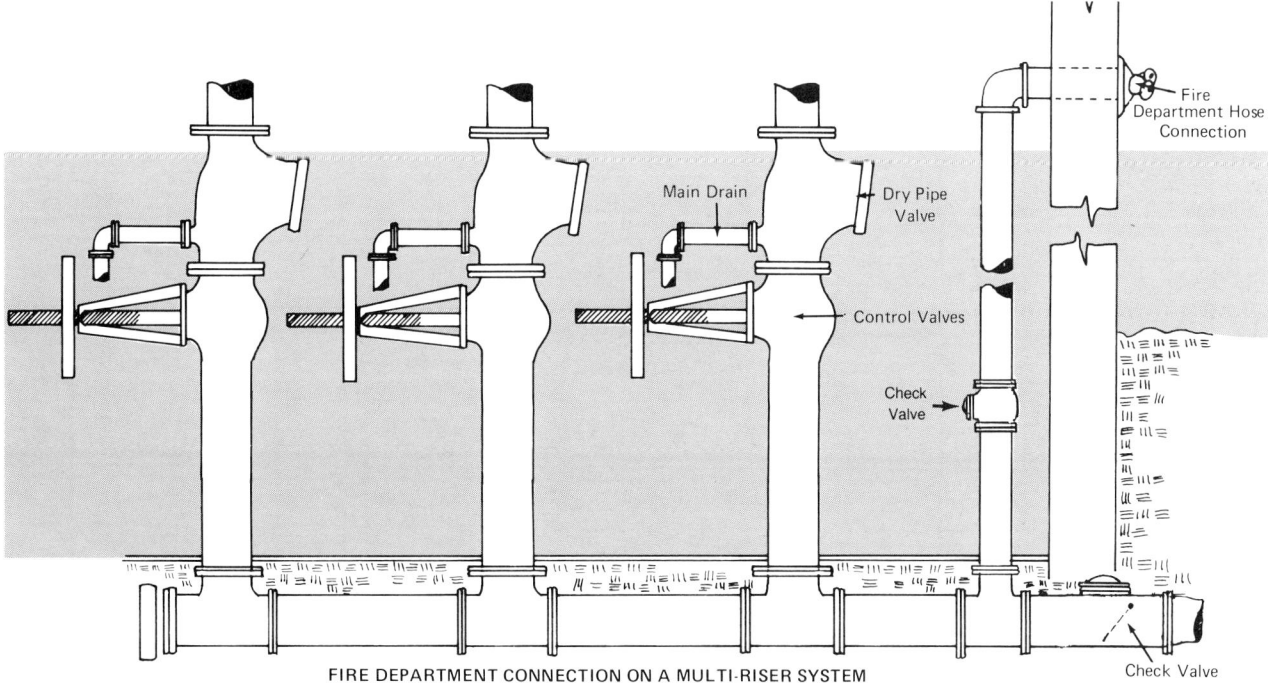

Figure 1-23 This piping schematic of a dry-pipe system illustrates that the fire department connection to the main riser is directly below the dry-pipe valve.

Figure 1-24 This piping schematic shows the position of the fire department connection for a multiriser dry-pipe system.

Most sprinkler systems are designed on the premise that a fire can be controlled by the operation of only a few sprinkler heads. If a large number of sprinklers are opened by the fire, the regular water supply may be inadequate and the flow from each sprinkler head may be so small that it is ineffective. Fire department pumpers operating from hydrants in the area may also further reduce the pressure in mains and rob the system of water. Whenever possible, fire department pumpers other than those supporting the system should operate from mains other than the primary water supply main for the system.

Fire department support of standpipes or automatic sprinkler systems should include a standard plan of operation. Develop 150 psi at the pumper and maintain this pressure if possible. Circumstances, such as those which exist in high-rise buildings or with deluge systems, may warrant a different pump discharge pressure. Such a plan cannot be established until fire department personnel become familiar with sprinklered properties under their jurisdiction. This should cover which buildings, the type of occupancy, type of system and the extent of the system. Therefore, an inspection survey is definitely a prerequisite. A thorough knowledge of the water system is likewise most important, including volume and pressure. These surveys should be diagrammed in a manner similar to the one illustrated in Figure 1-25. Fire department connections to sprinkler systems, the

Figure 1-25 Water supply diagrams should be drawn and kept with the operational plan for sprinkler-protected property.

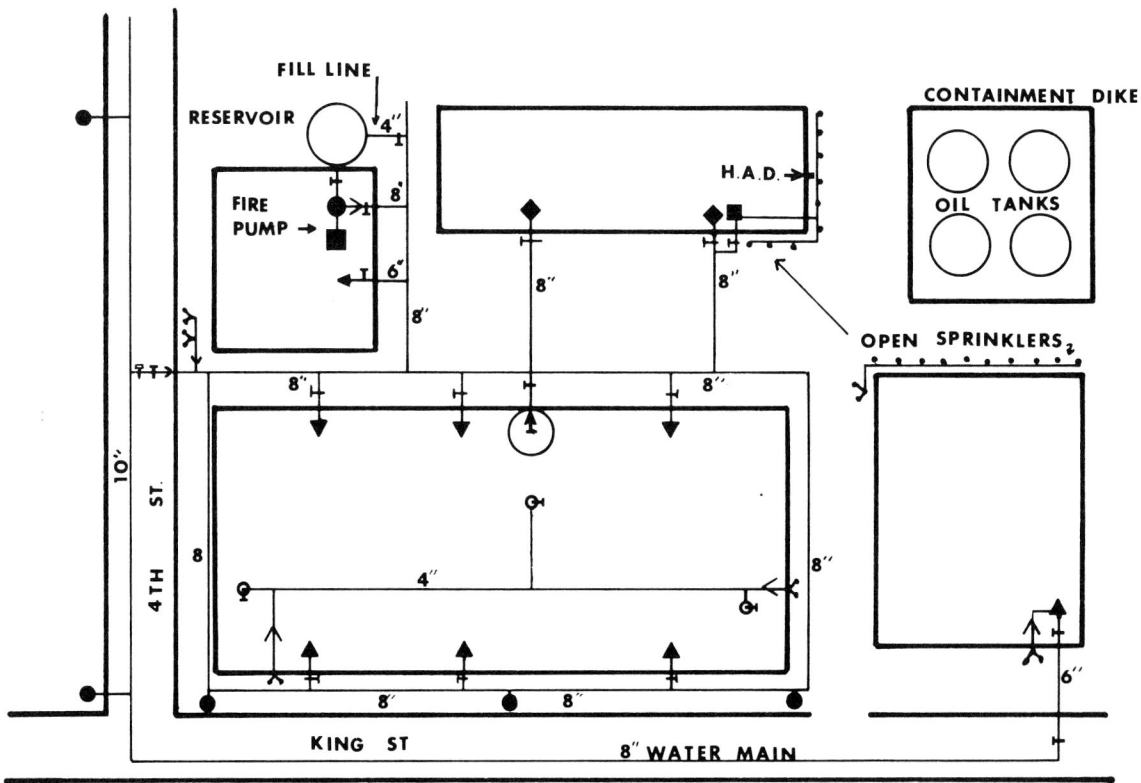

TYPICAL INSPECTION SURVEY DIAGRAM

Automatic Sprinkler Systems **25**

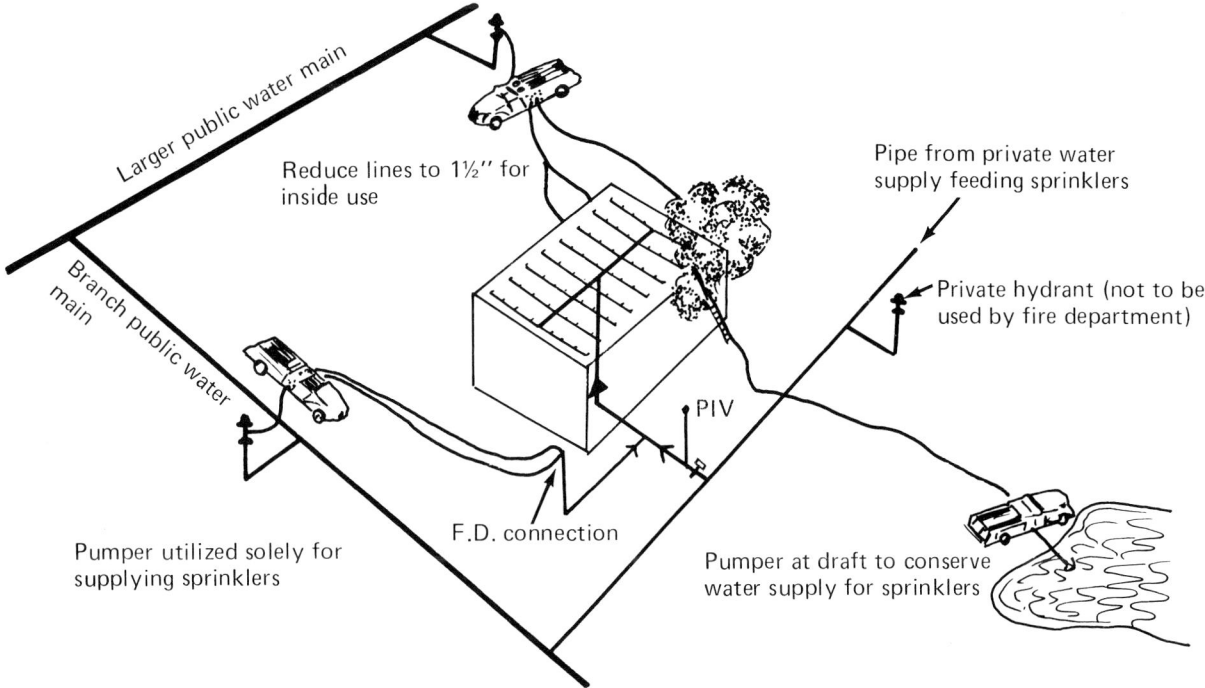

STANDARD PLAN OF OPERATION FOR FIRE DEPARTMENT SUPPORT OF AUTOMATIC SPRINKLER SYSTEMS

Figure 1-26 This diagram from the prefire plan indicats the location of fire department pumpers during an emergency. *(Courtesy of NFPA)*.

water supply available, the fire hydrants best suited for use to supply the system and the amount of hose necessary are most important. The fire department must be familiar with the area served by each connection. A plan should be followed once it has been established. Fire department pumper locations can be symbolized similar to the sketch illustrated in Figure 1-26. Some fireground support operation guidelines are listed:

- Pumper should be capable of supplying the necessary water and pressure.

- Use two hoselines (2½-inch or larger) from pumper to fire department sprinkler connections.

- A firefighter should immediately check the control valves to assure they are wide open. Where a fire pump is used, check to assure it is operating and valves are open. Open any closed valves. Carry a handlight and portable radio for communication.

- Fire officer-in-charge should locate the fire and order the hoselines charged. (The fire could be out.)

- The pump operator should slowly develop 150 psi at the pump and maintain this pressure if possible. If a third line is placed into the fire department connect, pump at 175 - 200 psi.

- Make a quick attack, entering the building with handlines.

- Perform ventilation as required.
- Get to unprotected areas or areas where sprinkler distribution might be blocked.
- Do not rob the sprinkler system of the water supply by overtaxing the water system with other pumpers.
- Support sprinkler systems in any exposed property.
- Avoid premature shutoff of a sprinkler system which has been in operation.
- Keep pumper and hoselines attached to sprinkler system during overhaul and station a firefighter at the valve to reopen if necessary.
- See that the sprinkler system is properly restored.
- During inspection, the condition of the fire department connection threads should be checked and protective caps replaced.

Further information can be obtained from NFPA Standard No. 13E, *Fire Department Operations in Properties Protected by Sprinkler and Standpipe Systems.*

THE WET-PIPE SYSTEM

Wet-pipe systems contain water under pressure at all times and are connected to the water supply so that a fused sprinkler head will immediately discharge a water spray in that area and actuate an alarm. This type of system is usually equipped with an alarm check valve that is installed in the main riser adjacent to where the feed main enters the building (Figure 1-27). This valve actuates an alarm when water flows through the system. Incor-

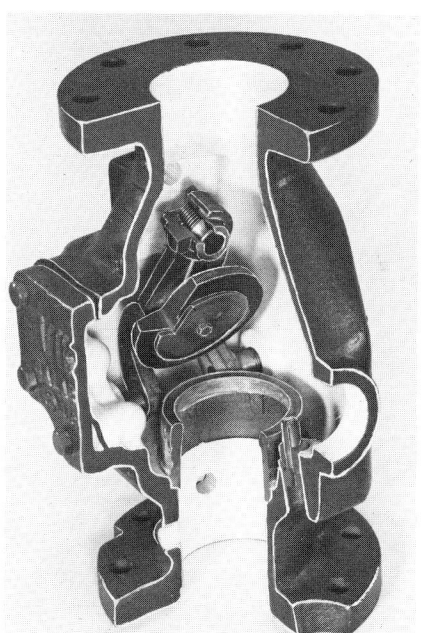

Figure 1-27 A sectional view of an alarm check valve showing the clapper and related parts. Note the auxiliary valve attached to the clapper.

porated in this valve is a clapper, which is simply an automatic one-way check valve with some additional features. Clappers are normally in the closed position. When a sprinkler head operates, this clapper opens and permits water to flow to the sprinkler and through the auxiliary valve to a retarding chamber to initiate an alarm signal.

Retarding Devices

The retarding chamber is installed between the alarm check valve and alarm-signaling equipment and is employed since water is subjected to variable pressures (Figure 1-28). This chamber is a time-delay device that retards the flow of water to the alarm equipment from the alarm check valve since it must be filled before the water can continue to the alarm equipment. Water drains through the small opening in the bottom of the chamber. If it were not for the retarding chamber, surges or increases in water pressure would cause the clapper in the alarm valve to rise momentarily and permit water to flow directly to the alarm equipment and thus transmit false alarms. A typical wet-pipe alarm check valve and retarding chamber in both the standby and fire positions are illustrated in Figure 1-29. This illustration will give a better idea of how the alarm check valve and retarding chamber operate.

Figure 1-28 The retard chamber's purpose is to prevent false alarms due to pressure surges in water supply.

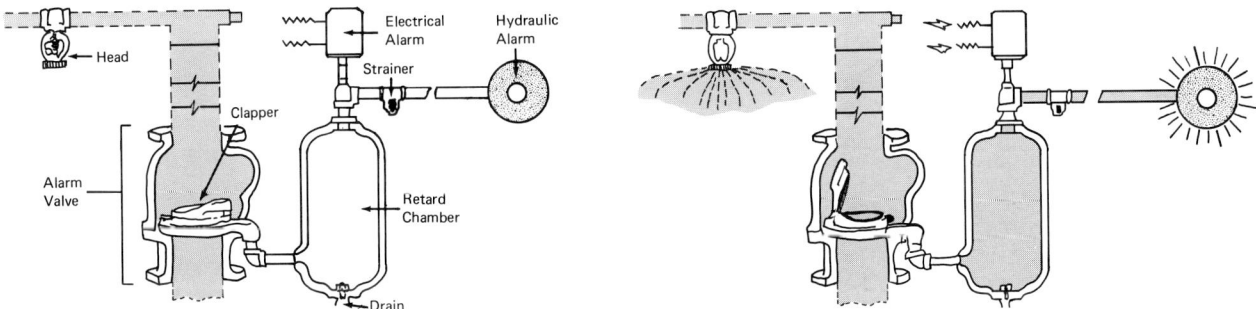

Figure 1-29 The retard system prevents normal pressure surges from tripping the alarm valve and sounding an alarm.

Another retarding device found where there are fluctuations in water pressure in pressure-maintenance pumps. These may be called excess-pressure pumps, jockey pumps or makeup pumps. These small capacity pumps are electrically operated. Their function is to reduce the running time of the main fire pump and keep it from starting each time there is a surge. With an initial pressure drop in the system, the jockey pump operates until its capacity is exceeded. The pressure in the system will continue to drop if there are sprinkler heads open. This will cause the sprinkler valve to open and the fire pump to operate, charging the system and shutting down the pressure-maintenance pumps.

28 PRIVATE FIRE PROTECTION

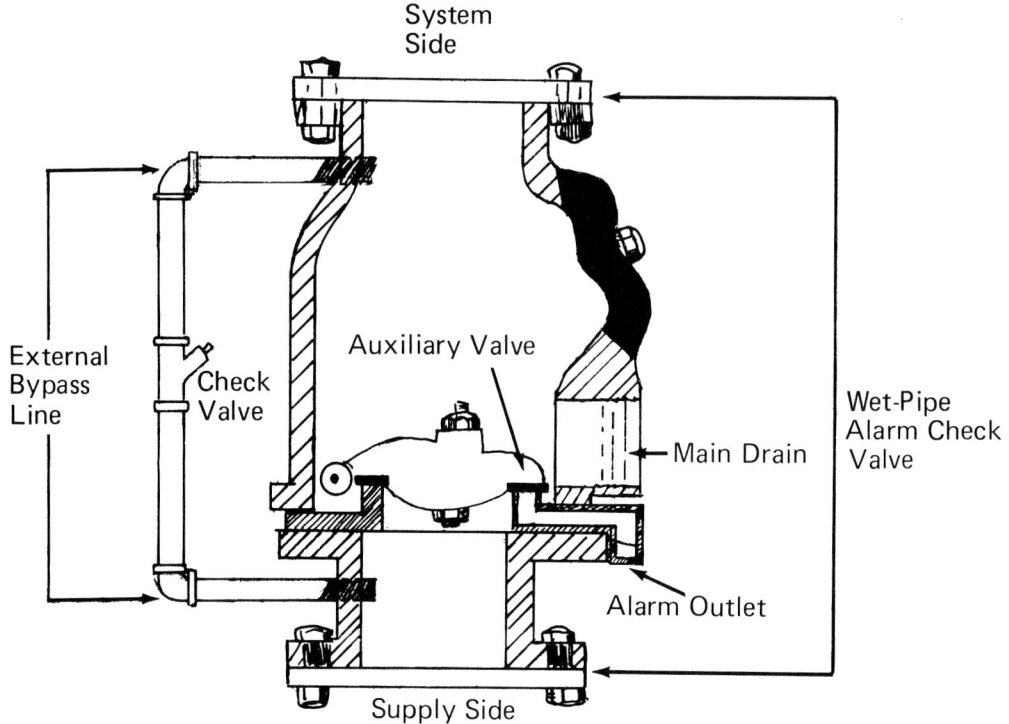

PIPING ARRANGEMENT FOR EXTERNAL BYPASS LINE

Figure 1-30 A typical alarm check valve utilizing an external bypass line to help prevent false alarms due to pressure surges.

Although surges or changes in water pressure are absorbed by the retarding chamber, they may, in addition, be offset by the use of an external bypass line with a check valve in the direction of flow. This is most common in areas with excessively high surges in water supply main pressure. By connecting this bypass line above and below the alarm check valve, water is permitted to flow around the clapper valve in one direction, from below (supply side) the clapper valve to above it (system side). The arrangement of the external bypass line is shown in Figure 1-30.

With surges of normal intensity in the riser, each successively higher pressure change passes into the system piping through the external bypass line around the clapper valve, which is the line of least resistance. It is then contained or trapped within the system through the function of the check valve in the external bypass unit and accumulates pressure higher than that normally supplying the system. The pressure gauges at the alarm check valve will record these pressures. In most sprinkler systems, two pressure gauges are installed; one above the clapper, which indicates pressure in the system, and one below the clapper, for an independent pressure reading on the supply riser. The accumulation of pressure in the system above the clapper valve, commonly referred to as excess pressure, is an important factor within any sprinkler system and is the basis for efficient and trouble-free serivce. So long as this excess pressure is maintained, the clapper valve will not normally raise from the seated position. Likewise, so long as this tightly seated position is maintained, water is withheld from the connecting piping to the retarding chamber and on to the signaling equipment.

Operation of the System

A diagram of a system is shown in Figure 1-31. When one or more sprinkler heads open, the flow of water lifts the main clapper off its seat and opens the auxiliary valve. The water pressure in the main riser pushes the clapper to full open position and continues to supply the open sprinklers. Water also enters the auxiliary valve alarm line and continues to fill the retarding chamber. When the retarding chamber is full, the water activates the pressure switch, sending an electrical alarm signal and a flow to activate the water-alarm gong. To shut down the system, turn off the main water control valve and open the main drain.

Figure 1-31 This is the flow of water in an alarm check valve and related parts during a fire condition.

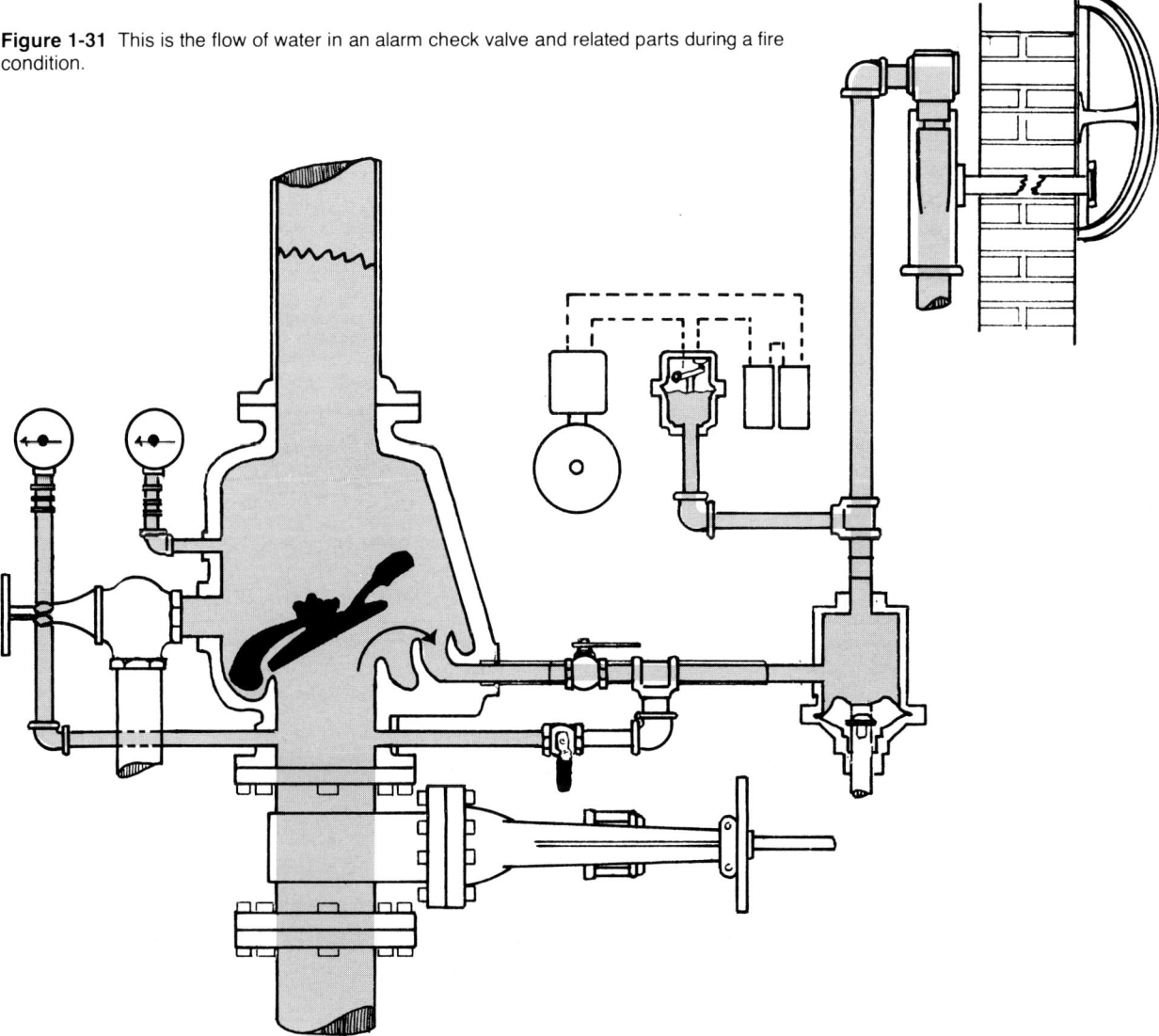

Using Sprinkler Stops

Until the officer-in-charge has determined that a fire is under control, the main sprinkler valve must not be closed. In order to decrease water damage, a sprinkler tong or wedge can be utilized to shut off the flow of water at the operating heads. Several types of sprinkler-head tongs and wedges can be carried in the pockets of turnout coats. Other types of tongs are attached to poles, which eliminates the need for a ladder. A wedge is made of soft wood which is designed to be driven into position with the heel of the hand until wedged between the sprinkler frame and water orifice. If the wedge is properly made, practically all the waterflow can be controlled. Sprinkler tongs are more effective in controlling the waterflow from a sprinkler because the rubber or neoprene stopper permits no dripping when applied properly. These tongs work on a clamp or lever principle as illustrated in Figure 1-32.

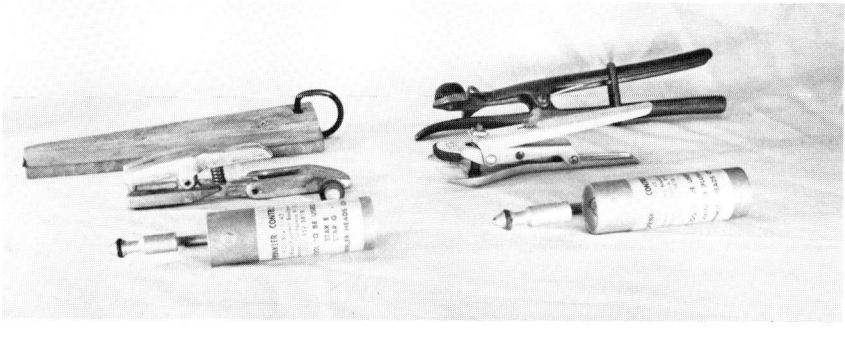
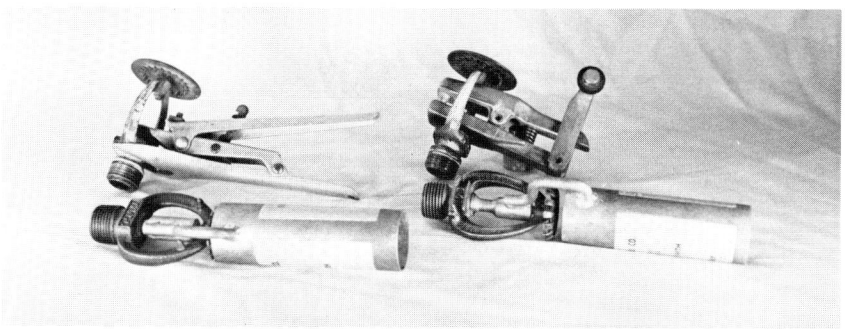

Figure 1-32 A variety of wedges and tongs that can be used to temporarily stop the flow of water from a fused head are available.

Restoring the System

Fire departments do not perform sprinkler maintenance

A sprinkler system should be installed in accordance with NFPA Standard No. 13. The installing contractor is required to provide the owner with instruction charts which describe the operation and proper maintenance of the system and NFPA Standard No. 13A on care and maintenance. Fire department personnel should not perform sprinkler maintenance and generally not restore a system which has been closed down. It is the responsibility of fire department personnel to see that the system is restored, but it is not their responsibility to restore it. It is wise for a chief officer to frequently recheck the fire scene until the automatic sprinkler system is back in service.

Inspecting the System

Sprinkler systems require periodic inspections and maintenance to perform properly during a fire situation. Firefighters may not be involved in regular testing of a system; however, they often are required to assist in putting a system back in service after a fire or accidental operation. The plant owner, maintenance personnel, fire prevention personnel and inspection personnel should be able to identify and inspect sprinkler systems when necessary. Routine maintenance should be performed by competent plant maintenance personnel or a contracted sprinkler company.

Several important steps should be taken prior to performing any inspection or maintenance on a sprinkler system.

- The inspector or maintenance person should review the records of prior inspections and identify the make, model, type and area of protection of the system.

- Permission from the plant management should be obtained before performing any inspection activity. The inspection should *never* be performed without this approval.

Obtain permission before initiating the inspection

- Firefighter, fire prevention or inspection personnel should *never* personally operate, adjust, physically manipulate, alter or handle any sprinkler devices or equipment during situations other than emergencies or planned training sessions. It is plant management's responsibility to provide or contract for personnel to perform these functions during normal conditions.

- If equipment is electronically supervised, prior to testing of sprinkler equipment, have plant personnel notify the alarm, monitoring organization.

- After all testing is complete, the plant personnel should inform the alarm monitoring organization that all testing has been performed. At that time, the alarm-monitoring organization should confirm that the alarm equipment did function as designed. If no alarms were received, the inspector should investigate the alarm equipment serviceability.

- Wear appropriate protective clothing for making a thorough inspection of attics, concealed spaces, and processing or manufacturing areas.

When a fire inspector or maintenance person is charged with the responsibility of performing inspections on wet sprinkler systems, his primary concern is with three major areas: 1) valves, 2) heads and 3) piping.

Inspect valves, heads and piping

VALVES

Inspect to assure that all valves controlling water supplies to the sprinkler system and within the system (sectional valves) are kept open at all times. When a valve is closed, report this

32 PRIVATE FIRE PROTECTION

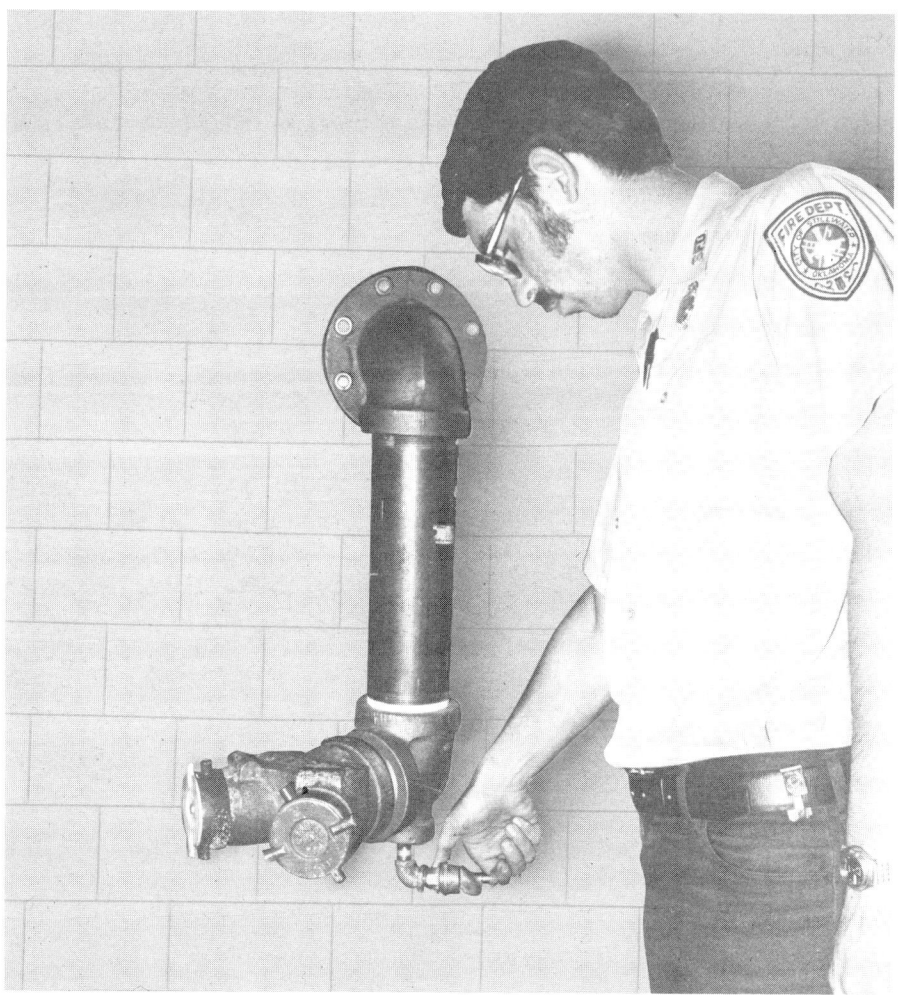

Figure 1-33 During an inspection, the operation of the ball drip valve on a fire department connection should be checked.

Figure 1-34 Check the velocity drip on the retard chamber to insure it is working properly.

condition to the responsible agency and fire department. Inspections common to all control valves are listed below. Examine each control valve for the following conditions:

- Valve is opened fully and secured and/or supervised properly in approved manner.
- Operating wheel is in good condition.
- Valve is accessible at all times. If a permanent ladder is provided, check the ladder to see that it is in good condition.
- Valve and its parts are not subjected to mechanical damage. Provide adequate guards if necessary.
- Wrench, when required, is in place.

In addition, inspect the post indicator valves (PIV) to see that they are fully opened. Try the wrench to feel the spring of the rod when the valve is fully opened. Keep the stem backed off about one-quarter turn from the full open position. See that the target is properly adjusted and cover glass is in place and clean. Assure that the PIV head bolts are tight and the barrel casing is intact.

Inspect the outside screw and yoke valves (OS&Y) to assure they are fully open, that the operating wheel is kept backed off approximately one-quarter turn and that the valve stem is clean.

Insure that all control valves are open

During normal conditions the inspector should insure:

- The alarm line shutoff cock is completely open.
- The pressure gauge valves are open.
- The static pressure above the clapper is equal to or greater than the static pressure below the clapper.

NOTE: Systems without check valves may only have one pressure gauge on the riser.

- The main drain valve, auxiliary drains and inspector's test valve are closed.
- The check ball drip valve in the fire department connection moves freely and allows trapped water to seep out (Figure 1-33).
- The velocity drip valve will move freely and allow trapped water to seep out of the retard chamber (Figure 1-34).

HEADS

It is important to make an immediate examination of sprinklers in areas where changes occur in occupancy, fire hazard, heating, lighting or mechanical equipment since these changes may require the installation of different types of sprinkler heads. Inspect all sprinkler heads to make sure they are clean, not damaged and free from corrosion. The need for guards for protection against mechanical injury should be reported. Sprinkler heads in buildings subject to high temperatures should be carefully examined. Any head showing evidence of weakness should be replaced with heads of the proper temperature rating. Weak heads are indicated by a creeping or sliding apart of the fused parts, which is known as cold flow, or by leakage around the head. Cold flow is caused by repeated heating of a head to near its operating temperature. Cold flow problems can be eliminated by the use of frangible bulb heads. Sprinkler heads exposed to a corrosive atmosphere should have a special protective coating. Heads that are badly corroded, painted or heavily loaded with foreign material should be replaced.

Watch for obstructions to sprinkler head discharge

Partitions or stock should not obstruct the distribution of water discharge from sprinklers and sprinkler heads should be free of hanging displays. A clearance of at least 18 inches should be maintained under sprinklers, measured from the deflector. An adequate supply of sprinkler heads should be kept in a cabinet together with a sprinkler wrench. The extra supply should include sprinkler heads of the various designs and temperature ratings in service so that prompt replacement of used or damaged heads can be made and full protection restored. Sprinklers for hallways, shafts and special rooms may have special deflectors.

PIPING

Inspect all sprinkler piping and hangers to determine that they are in good condition. A check should be made for corrosion and areas which could be subject to physical damage, assuring there are no leaks. Sprinkler piping is not to be used as a support for ladders, stock or other material. Report all bent or damaged pipes and missing hangers. Report all loose hangers, as they may put an unnecessary strain on the piping and fittings, cause breaks and interfere with proper drainage (Figure 1-35).

Piping in wet systems must be protected against freezing. Freezing can stop the flow of water to sprinklers or cause the failure of control and alarm devices. Also, the piping may be ruptured, causing severe water damage or expensive repairs and interruption of protection.

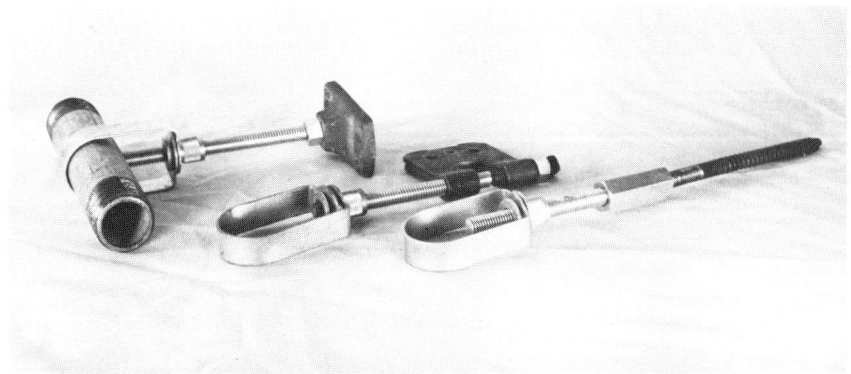

Figure 1-35 This is an example of the many types of branch-line hangers.

Wet-Pipe System Testing

ALARM TEST

On a wet-pipe system with an alarm check valve, open the alarm bypass valve to test the alarm without unseating the valve (Figure 1-36). The pressure gauge readings should not change significantly, but the water should flow to the retard chamber (if so equipped), and then to the alarm line. The water motor gong or electric alarm should sound. The retard chamber drain should empty the chamber after the alarm bypass valve is closed. If there is no retard chamber, the alarm line should be drained at the conclusion of the test.

WATERFLOW ALARM TEST

On all types of wet systems, an alarm test should be conducted by using the inspector's test connection (Figure 1-37). With an observer at the riser, another individual at the inspector's test connection, which should be located at the farthest end of the system, opens the inspector's test valve. Water should flow from the exterior orifice at the test connection. Observe the pressure gauge(s). Only a slight variation in pressure should be observed, but the alarm should sound. After hearing the alarm, the inspector's test valve should be closed.

Figure 1-36 To test alarm devices without unseating the alarm check valve, open the alarm bypass valve.

Figure 1-37 To test alarm devices through the alarm check valve, open the inspector's test valve until the alarm operates.

MAIN DRAIN TEST

Step 1: Observe and record the pressure on the gauge(s) at the system riser (Figure 1-38).

Step 2: Fully open the 2-inch main drain, observe and record the pressure drop as the alarm sounds (Figure 1-39).

Step 3: Close the 2-inch main drain and compare readings to previously recorded readings (Figure 1-40). If significant differences are noted, a supply valve may be partially closed or an obstruction may be present in the supply line. On a system using an alarm check valve, take the pressure readings from the lower gauge, since erroneously high static pressures can exist above the valve.

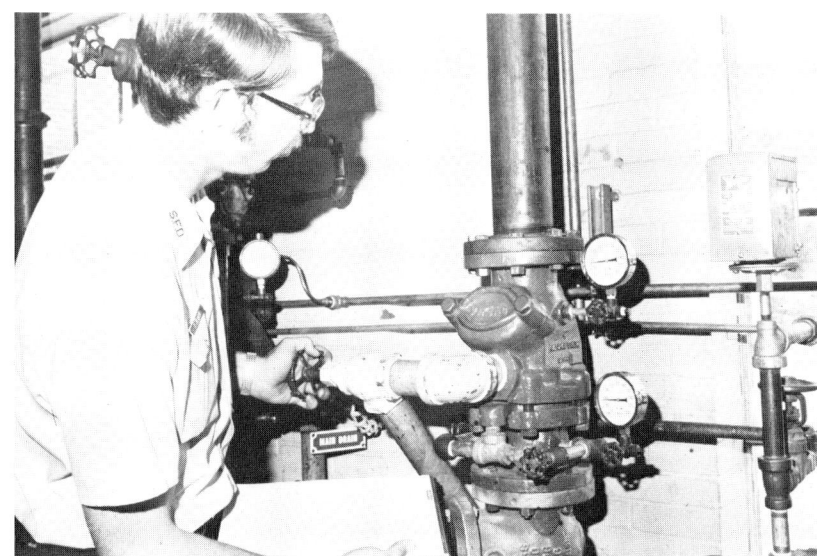

Figure 1-38 Start a main drain test by recording pressure readings at the riser.

Figure 1-39 To conduct the main drain test, open the 2-inch drain valve and record any pressure drop as the alarm operates.

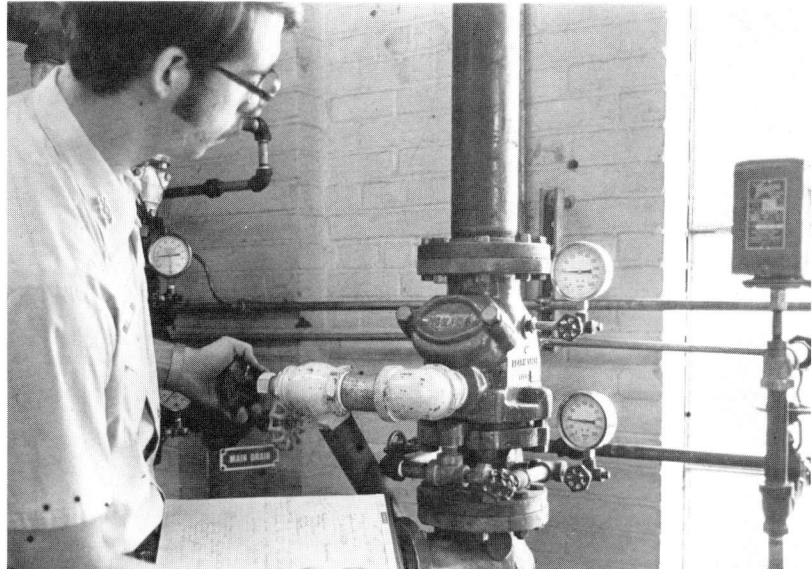

Figure 1-40 Return the 2-inch drain to its normal closed position to complet the test, and compare the recorded gauge readings.

Frequency of Inspecting and Testing Wet-Pipe Sprinkler System

Wet-pipe sprinkler systems should be visually inspected at least monthly; however, a weekly visual inspection is recommended since it provides the in-plant inspector with continual familiarization with the system and an early discovery of system deficiencies.

Alarm test and waterflow alarm testing should be accomplished only during nonfreezing weather to avoid ice formation on sidewalks and roadways. Also, the alarm piping can be damaged by cold-weather testing. These tests should be performed at least yearly, but monthly or even weekly tests are recommended. During tests, all the visual items should also be checked.

The 2-inch main drain test should be conducted at least annually; however, a quarterly test is suggested. The inspector should visually inspect the system and perform the alarm test in conjunction with the drain test. A written record should be kept on file after each test.

DRY-PIPE SPRINKLER SYSTEM

A dry-pipe sprinkler system is one where air under pressure replaces water in the sprinkler piping above the dry-pipe valve. A dry-pipe valve is a device that keeps water out of the sprinkler piping until a fire actuates a head or heads. Dry systems should be used in buildings where insufficient heat is maintained to keep water from freezing. When a sprinkler head fuses, permitting the air pressure to escape, the dry-pipe valve automatically opens to permit water to replace the air pressure in the lines. Dry-pipe valves are designed so that a small amount of air pressure above the dry-pipe valve in the sprinkler piping will hold back a much greater water pressure on the water supply side of the dry-pipe valve. Dry systems are equipped with either electric or hydraulic alarm-signaling equipment. Figure 1-41 illustrates the dry-pipe valve in both the standby and fire positions.

The two types of dry-pipe valves that are in use today are the differential and the mechanical. Study Figure 1-42 for the clarification of the operation of the differential dry-pipe valve. The differential type has a double-seated valve or two clappers of unequal size. The upper air seat is considerably larger than the lower, or water, seat with a surface ratio that is normally six to one. The difference in surface area of these two seats determines the difference between water pressure and air pressure that is necessary to "balance" the valve.

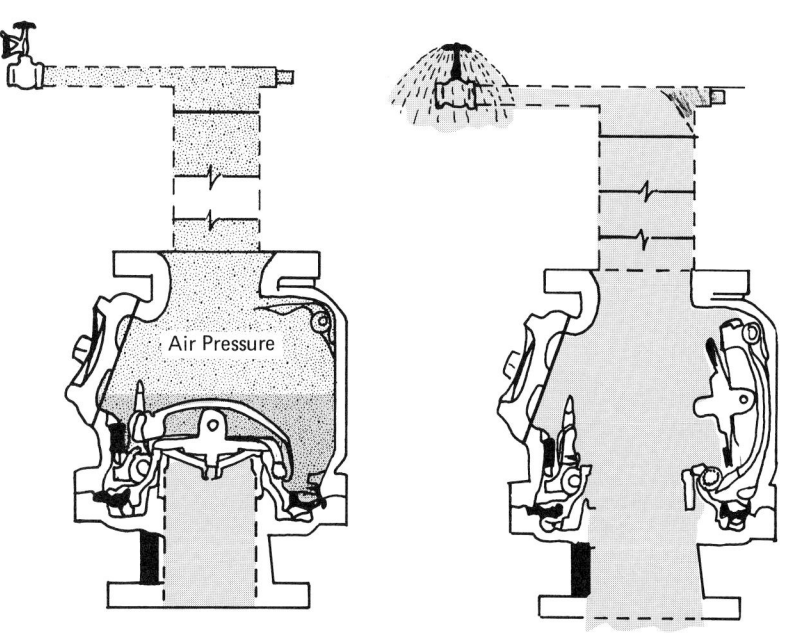

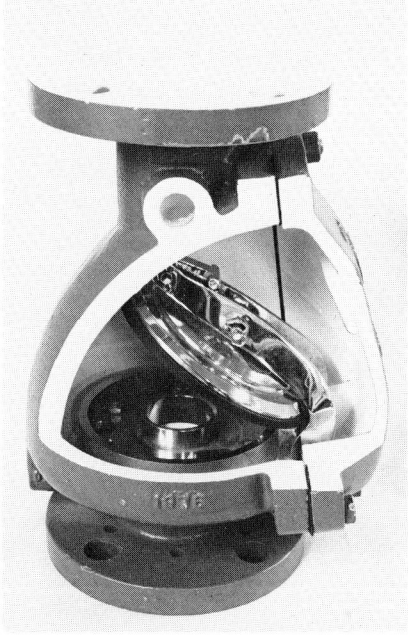

Figure 1-42 A differential dry-pipe valve obtains its name from the small water seat in the center of the valve as compared to the large air seat above it.

Figure 1-41 The dry-pipe alarm valve is shown in the standby position and in fire position. Notice the prime water used to make a seal between the clapper and seat when the valve is in the standby position.

This difference can be seen as follows: The air/water differential serves a two-fold purpose. First, a comparatively small amount of air pressure is required to hold the system operational. Secondly, the less air pressure throughout the system, the more quickly this air can be expelled by water when a sprinkler head fuses; thus, the time needed for water to be discharged is reduced.

Air pressure in the system should not be permitted to vary beyond the prescribed limits. High air pressure is difficult to maintain and operation of the valve is considerably delayed if excessively high pressure is used. In systems where pressure from the primary water supply is low and where a fire pump is installed, sufficient air pressure must be maintained to keep the dry-pipe valve from accidentally tripping when the fire pump operates causing high water pressure.

Low-differential dry-pipe valves resemble alarm or check valves. These single clapper valves depend upon air pressure in the system piping of approximately 110 percent of the maximum static water pressure to keep the dry-pipe valve closed against the water. The low differential valve has a differential air to water ratio of 1.1 to 1 as opposed to the traditional 6 to 1 ratio. They are equipped with a pilot valve or split seating to provide a fire alarm signal upon operation of the system. The early development of the low-differential valve resulted because dry-pipe systems were being damaged by debris carried by high-velocity water rushing through the dry-pipe valve into the sprinkler system. This problem was typical of systems taking water from nonpotable water sources. The answer appeared to be to more closely equalize the pressure differential of the valve. The velocity of the water was reduced, and correspondingly the debris introduction was reduced.

The advantages of the valve are: low-velocity water entry, a valve trip time 85 percent quicker than a traditional valve, a total flow time reduction of nearly 70 percent over a traditional valve, they are easy to reset and they can be utilized as either wet or dry valves. The disadvantages are that a heavier-duty air compressor is needed, an automatic pressure-maintenance device is required and protection against a water column above the clapper is required.

In the mechanical type of dry-pipe valve, the air and water seats may or may not be of the same area. The air seat (air clapper) under the influence of air pressure holds the water seat (water clapper) shut through a system of multiplying levers. Another type of mechanical dry-pipe valve has only one clapper that is held closed by a system of levers. The levers and clapper are released by a difference in air pressure acting on a diaphragm. These valves are obsolete and no longer manufactured. A standard recommendation would be to have the valve replaced with a modern valve.

The required air pressure for dry systems usually ranges between 15 and 50 psi. Air pressure that is needed to service a dry system may be derived from two different sources. These sources are either from plant air service or from unit air pressure that is supplied by a compressor and tank used exclusively for the sprinkler system.

Low-differential dry-pipe valves reduce flow time by 70 percent

Accelerators and Exhausters

In a large dry system several minutes could be lost while the air is being expelled from the system. Rules have been established which normally require a quick-opening device to be installed in systems that have a water capacity of over 500 gallons. There are two types of quick-opening devices: accelerators and exhausters. The accelerator unbalances the differential in the dry-pipe valve, causing it to trip more quickly, whereas the exhauster quickly expels the air from the system. When a sprinkler head is fused and air pressure in the accelerator-type system drops a few psi (usually one or two pounds), a diaphragm in the accelerator becomes unbalanced (Figure 1-43). This unbalanced condition causes a valve to open, which permits the air pressure in the system to enter the intermediate chamber of the dry-pipe valve. As soon as air is equalized on both sides of the air clapper (normally 10 to 15 seconds), the valve is automatically tripped by water pressure. In the exhauster type (Figure 1-44) the fusing of a sprinkler head causes a diaphragm to open a large valve. This action permits air pressure to quickly escape to the outside and the dry-pipe valve to trip. Both of these devices are complicated mechanisms, and they require proper care and maintenance. They should be tested in the spring and fall by a competent individual. Although it will take longer, the dry-pipe valve will operate, even if the quick-opening devices do not operate.

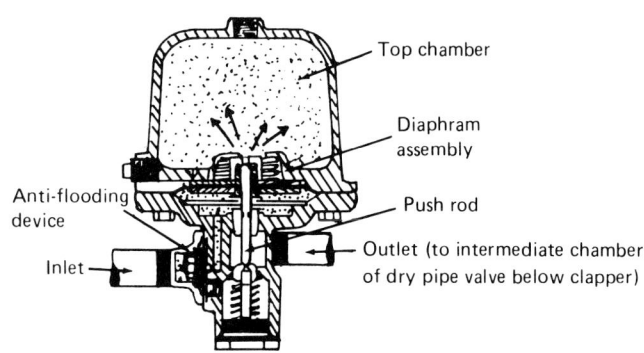

A DRY PIPE ACCELERATOR

Figure 1-43 When activated by a reduction in pressure, a dry-pipe accelerator channels air pressure to the intermediate chamber directly under the clapper valve, thus forcing it open. *(Courtesy of Reliable Automatic Sprinkler Co., Inc.)*

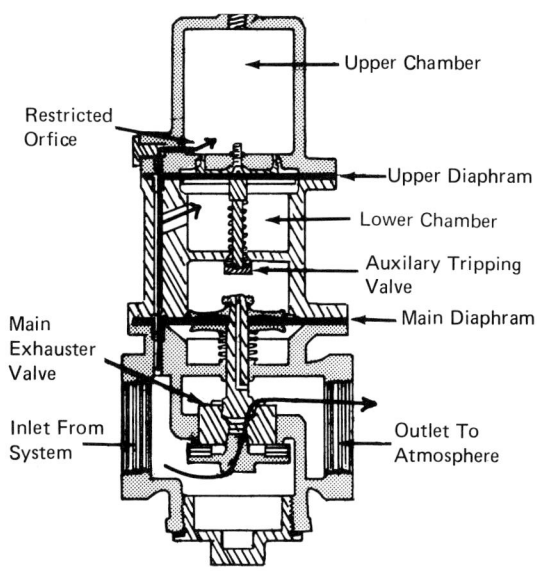

A DRY PIPE EXHAUSTER

Figure 1-44 When activated by a reduction in air pressure, a dry-pipe exhauster discharges existing system air pressure to the atmosphere. *(Courtesy of Reliable Automatic Sprinkler Co., Inc.)*

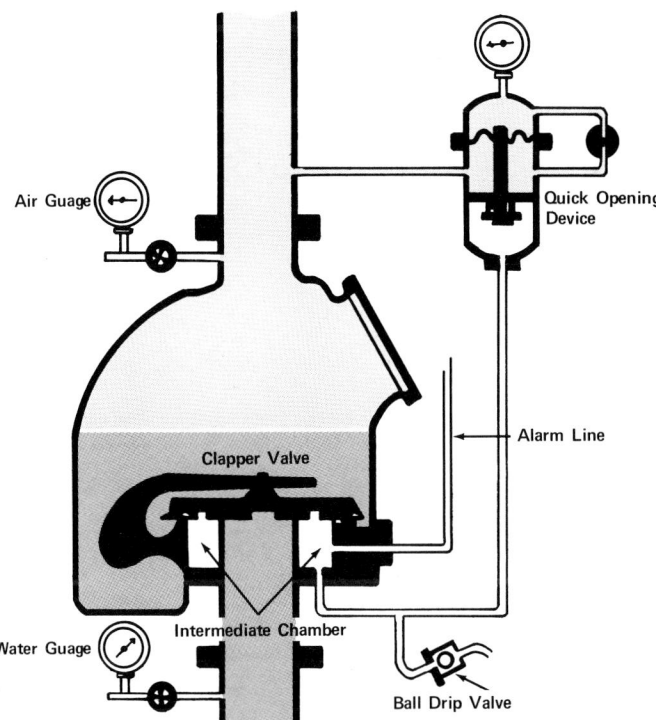

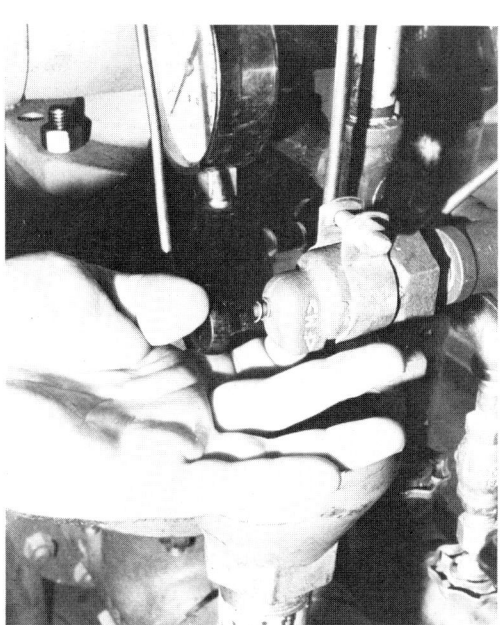

Figure 1-45 The component parts of a dry-pipe valve are illustrated to show the system's operation.

Figure 1-46 Inspect the drip check valve to insure proper drainage.

Operation of the System

Figure 1-45 shows the component parts of the dry-pipe system. When one or more sprinklers open, the air pressure is vented through the open heads from the system, thus upsetting the differential within the dry-pipe valve. An accelerator or exhauster will be an aid in the speeding-up operation of the system. As the differential is upset, the riser water pressure raises the clapper assembly into the wide-open locked position. As the water fills the upper chamber of the dry-pipe valve, it also enters the intermediate chamber where it forces the drip check valve closed. It then flows into the alarm line to activate the alarm equipment. The water continues to the open sprinklers.

Restoring the System

To shut down the system, as with the wet system, the main water control valve must be closed and the main drain opened. The air supply for the system should now be shut off. It may also be necessary, depending on piping arrangement, to drain any low points in the system, including the alarm line and alarm bypass test line, by opening their valves. Also, examine the drip check valve to make sure it has opened and drained when pressure was reduced (Figure 1-46).

The fire department is not responsible for maintaining or restoring a dry-pipe system, although it is the responsibility of fire department personnel to see that the system is restored. It is wise for a chief officer to frequently check the premises until the automatic sprinkler system is back in service.

Figure 1-47A This photo shows the position of the ball drip valve on a fire department connection.

Figure 1-47B A dissassembled ball drip valve showing the check ball and related passageways.

Inspecting the System

During a normal inspection, the inspector should insure that:

- All indicating control valves are open and properly supervised in the open position.
- Air pressure readings correspond to previously recorded readings.
- The ball drip valve will move freely and allow trapped water to seep out of the fire department connection (Figure 1-47).
- The velocity drip valve located beneath the intermediate chamber is free to move and allow trapped water to seep out (Figure 1-48). This valve can be checked by lifting a push rod which extends through the drip valve opening. Where an automatic drip valve is installed, the valve can be checked by using the push rod located in the valve opening.
- The fire department connection threads are in good condition and the caps are in place.
- Any drum drips are drained to relieve the moisture trapped in the low areas of the system (Figure 1-49).
- During freezing weather, check the dry-pipe valve-enclosure heating device at frequent intervals to insure the temperature is maintained at or above 40°F.
- The priming water is at the correct level; drain if necessary by opening the priming water test-level valve until air begins to escape (Figure 1-50).

  NOTE: If the system is equipped with a quick-opening device, opening the priming water test line could trip the system.

- The system's air pressure is maintained at 15 - 20 psi above the trip point and no air leaks are indicated by rapid or steady air

Automatic Sprinkler Systems **43**

Figure 1-48 The velocity drip valve drains the intermediate chamber. Note the push rod.

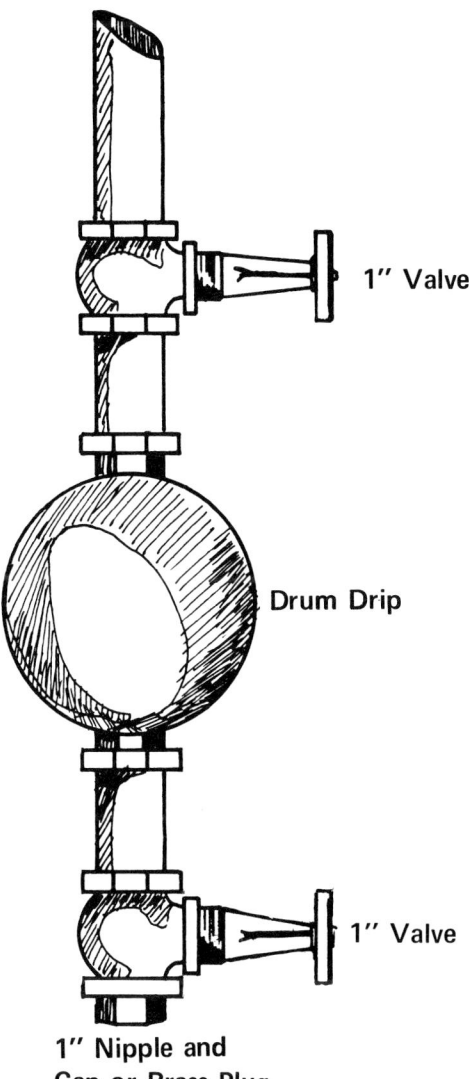

Figure 1-49 A drum drip valve for removing moisture from the system piping.

Figure 1-50 Drain off excess water from the priming chamber to obtain the correct priming-water level.

loss. If excessive air pressure is noted, it can be drained using the priming-water test-level valve; however, if the system is equipped with a quick-opening device, it is necessary to shut off the device to prevent tripping the valve. Close the quick-opening-device control valve and open the priming-water test-level valve to reduce the air pressure. Close the valve when the prescribed limit is reached. Before opening the quick-opening-device control valve, reduce the air pressure in the quick-

44 PRIVATE FIRE PROTECTION

Figure 1-51 Record the static pressure before a main drain test.

Figure 1-52 Open the main drain and check for water flow.

Figure 1-53 Record the flow pressure.

Figure 1-54 Close the 2-inch drain slowly to avoid a water hammer which could damage the system.

opening device until it reaches the pressure in the system. In cases where there is no provision made for bleeding air from the quick-opening device, the air pressure may be reduced by loosening the air gauge.

Dry-Pipe System Testing

MAIN DRAIN TEST

Step 1: Record the static pressure (Figure 1-51).

Step 2: Fully open the 2-inch main drain slowly to check the water flow (Figure 1-52).

NOTE: Open and close the 2-inch main drain very slowly to avoid tripping the dry-pipe valve due to water hammer.

Step 3: Record the flow pressure (Figure 1-53).

Step 4: Close the 2-inch main drain slowly (Figure 1-54).

Step 5: Test the alarm by opening the alarm test bypass valve (Figure 1-55).

NOTE: The line from the intermediate chamber should be equipped with a check valve and a globe valve (Figure 1-56) to prevent water from flowing back into the intermediate chamber and tripping the valve. If this is the case, close the globe valve before opening the bypass valve to prevent tripping in case the check valve malfunctions.

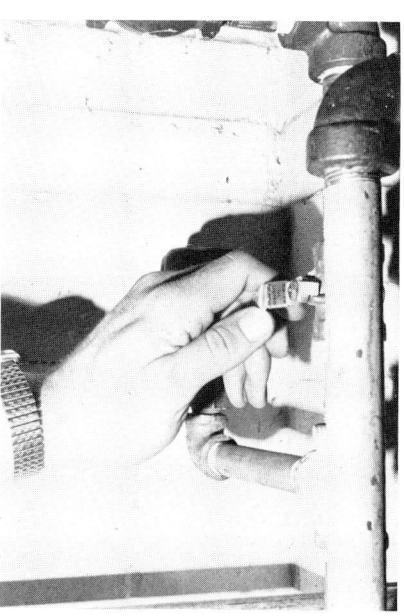

Figure 1-55 Open the alarm test bypass valve to test the alarm.

Figure 1-56 A system equipped with a check valve and globe valve to prevent a back flow of water into the intermediate chamber.

TRIP TEST

Step 1: Check the system control valve (OS&Y) for sticking by opening the 2-inch main drain slightly while closing and reopening the supply control valve (Figure 1-57).

Step 2: Close the 2-inch main drain (Figure 1-58). The 2-inch main drain is used in this case to prevent accidental tripping. Closing the control valve without opening the 2-inch main drain can "squeeze" the water trapped between the control valve and the dry-pipe clapper. Since water is incompressible, for all practical purposes, something has to give — and in many cases the clapper will be pushed off its seat.

Step 3: Assign one person to open the inspector's test valve and record the time required before water discharges (Figure 1-59).

Step 4: Record tripping point of the dry-pipe valve as indicated by the air pressure gauge (Figure 1-60).

Step 5: Check water and air pressure gauges to insure pressure equalization has occurred after tripping (Figure 1-61).

Figure 1-57 Check the main control valve before initiating a trip test.

Figure 1-58 Close the main drain after checking the supply control valve.

Figure 1-59 Operate the inspector's test valve and record the time to discharge water.

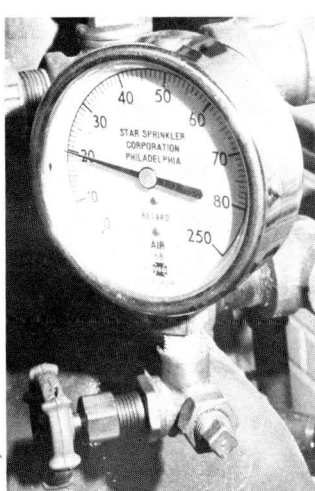

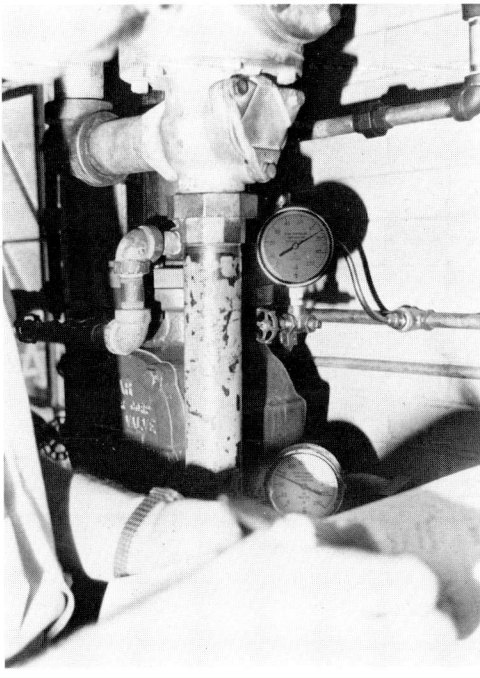

Figure 1-60 Personnel must watch the air pressure gauge to record the air pressure when valve trips.

Figure 1-61 Check for water and air pressure equalization after the valve has tripped.

Step 6: Close the supply control valve (OS&Y) and open the main drain valve (Figure 1-62). Be sure that the system is completely drained before proceeding. Accumulation of a column of water above the clapper (water-columning) can prevent valve operation.

Step 7: Close the inspector's test valve after the system is drained (Figure 1-63).

Step 8: Open the dry-pipe valve and check to make sure the clapper is latched in the open position (Figure 1-64).

Step 9: Clean the clapper seat and any debris from the valve housing, close valve face after reseating the clapper (Figure 1-65).

Figure 1-62 Open the main drain valve to drain the system.

Figure 1-63 Close the inspector's test connection after the system is drained.

Figure 1-64 Open the dry-pipe valve cover and check that the clapper is latched in the open position.

Figure 1-65 Clean the clapper seat, remove debris and replace the valve cover.

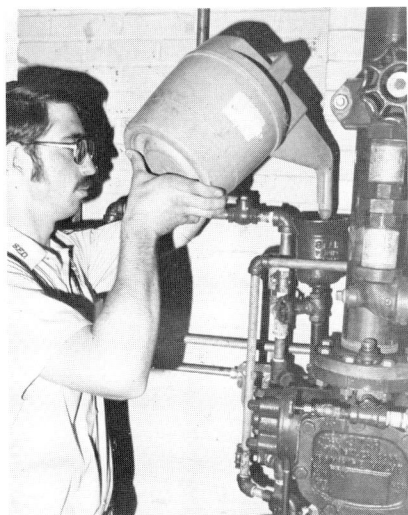

Figure 1-66 Fill the valve with priming water to the appropriate level.

Figure 1-67 Pressurize the system to the proper air level.

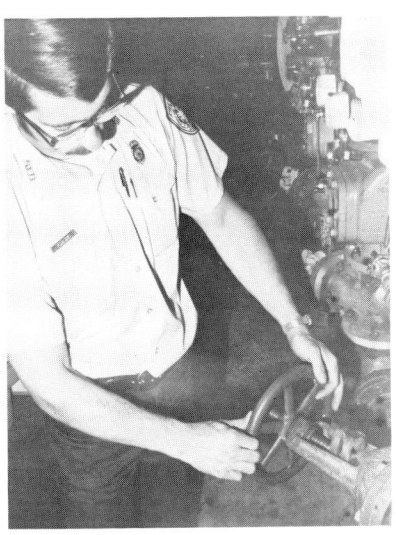

Figure 1-68 Open the main drain valve while opening the supply control valve to restore water pressure.

Figure 1-69 Close the main drain slowly after opening the supply control valve fully.

Step 10: Add priming water until the priming-water level indicator flows water (Figure 1-66).

Step 11: Pressurize the system with air to the proper level (Figure 1-67).

Step 12: Restore water pressure by opening the supply control valve with the 2-inch main drain open to avoid tripping the valve (Figure 1-68).

Step 13: When the supply control valve is completely open, slowly close the 2-inch main drain (Figure 1-69). Check gauges.

NOTE: This procedure can take 2 to 4 hours to complete, depending upon the amount and size of the piping in the system and the capacity of the air compressor to pressurize the system.

Frequency of Inspecting and Testing of Dry-Pipe Systems

Dry-pipe sprinkler systems should be visually inspected at least weekly. A weekly inspection is recommended since it provides the in-plant inspector with further knowledge of the system and an early discovery of system deficiencies. This is particularly important with respect to discovering air leaks.

The 2-inch main drain and alarm test should be performed only during nonfreezing weather to avoid alarm piping damage and ice accumulations causing safety hazards. The 2-inch main drain test should be performed at least annually; however, a quarterly test is suggested, if conditions permit. An alarm test should be performed at least monthly although a weekly test is recommended. During these tests a visual inspection for abnormal conditions should be performed.

The trip test should be accomplished at least annually. The inspector should visually inspect the system and perform the drain test in conjunction with the trip test. A written record of all tests and inspections should be kept on file for comparison later.

PRE-ACTION SYSTEM

A pre-action system is a dry system which employs a deluge-type valve, fire detection devices and closed sprinkler heads. This type of system is used when it is especially important that water damage be prevented, even if pipes should be broken. The system will not discharge water into the sprinkler piping except in response to the detection system. A system which contains over 20 heads must be supervised so that should the detection system fail the system would operate automatically.

Pre-action systems are used to prevent water damage

An example of a pre-action system is shown in Figure 1-70. Fire detection and operation of the system introduce water into the distribution piping prior to the opening of any sprinklers. In this system, fire detection devices operate a release located in the system-actuation unit. This release opens the deluge valve and permits water to enter the distribution system so that water is ready when the sprinkler heads fuse. When water enters the system, an alarm sounds to give a warning prior to the opening of the sprinkler head.

Inspecting and testing the system will be essentially the same as that for a deluge system. See "Inspection and Tests" under "Deluge Sprinkler System" on Page 52.

DELUGE SPRINKLER SYSTEM

This system is ordinarily equipped with open sprinkler heads and a deluge valve. Fire detection devices are installed in the same area as the sprinkler heads. Upon fire detection, the deluge valve is opened, which permits water to flow into the system and out of **all** the sprinkler heads. The purpose of a deluge system is to

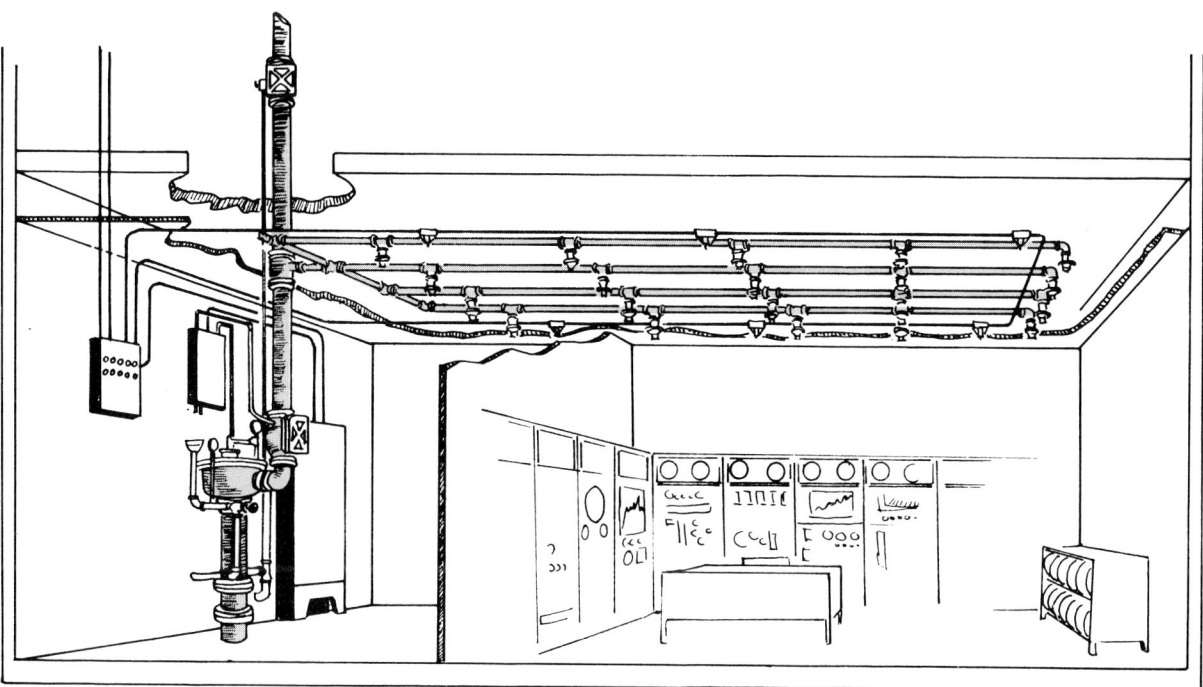

Figure 1-70 This pre-action system with detection devices, controls, closed sprinkler heads and a deluge valve begins to operate before the sprinkler heads open.

wet down the area in which a fire originates by discharging water from all open heads in the system. This system is normally used to protect extra-hazardous occupancies. Many modern aircraft hangars are equipped with an automatic deluge system which may be combined with an automatic sprinkler system. A system using partly open and partly closed heads is considered a variation of the deluge system.

Operation of the System

Activation of the deluge system may be controlled by fire, heat or smoke detecting devices and a manual device. One type of deluge valve in both the standby and fire position is illustrated in Figure 1-71. Since the deluge system is designed to operate automatically and the sprinkler heads and nozzles do not have heat-responsive elements, it is necessary to provide a separate detection system. This detection system is connected to a tripping device which is responsible for activating the system. As there are several different modes of detection there are also many different methods of operating the deluge valve. Deluge valves may be operated electrically, pneumatically or hydraulically.

An electrically operated deluge valve is designed for use with a fire detection system which transmits an electrical signal to the valve. The activation system will include an electrical tripping mechanism which will electrically release the deluge valve clap-

Figure 1-71A A pre-action system in standby position.

Figure 1-71B A pre-action system in fire position.

per. Also included will be a manual release and reset in case of power failure. There will be a primary power source such as normal line voltage; however, batteries are usually provided as a secondary power source. A major advantage of the electrically operated deluge valve is its speed of operation.

Pneumatically operated deluge valves are designed to be used with a pneumatic detection system, usually rate-of-rise detectors. This type of activation is actually a combination of pneumatic detection and mechanical activation. A pneumatic detector which is discussed in Section 4 operates on an imbalance of pressure. Activation of the deluge valve occurs when this change in air pressure displaces a diaphragm which is mechanically linked to a tripping mechanism that in turn releases the deluge valve clapper. The pneumatic system does not require electrical power for activation; however, a manual release is required.

There are several different types of hydraulically operated deluge valves. Some of these valves may utilize a combination of a hydraulically operated valve and a dissimilar type of detection system such as pneumatic, hydraulic or electric. Usually, no matter what combination is used, hydraulic pressure is depended on for activation of the system by changing the deluge-valve differential. Examples of hydraulically operated valves are shown in Figure 1-72.

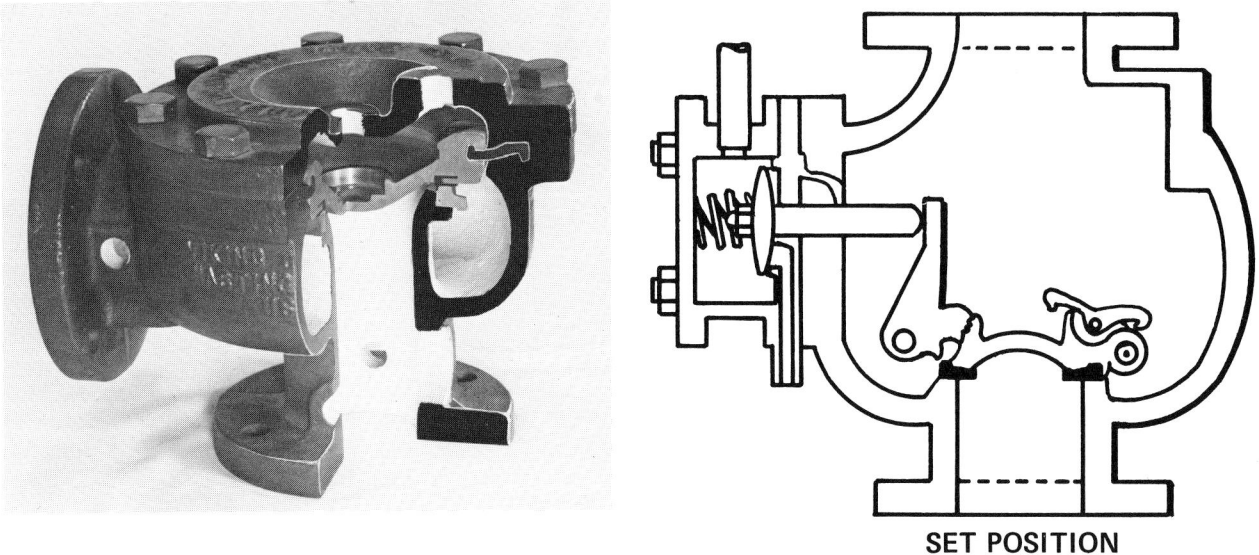

Figure 1-72 Two hydraulically opened deluge valves. The one on the right has a positive-latch feature and can be activated by a detection system.

When the system is actuated by the automatic detection system or a manual station, the water pressure in Chamber A is immediately lowered to almost zero. This destroys the differential between Chamber A and Chamber B. The component parts involved in this operation are shown in Figures 1-73 and 1-74. This pressure drop will be indicated on the pressure gauge for Chamber A. The pressure in Chamber B then forces the clapper valve upward, allowing water to enter Chamber C and to continue on into the system piping and sprinkler heads. Some of the water entering Chamber C will flow through the indicator valve and piping to activate the alarm equipment. To shut down any system, close the main water control valve and open the main drain valve.

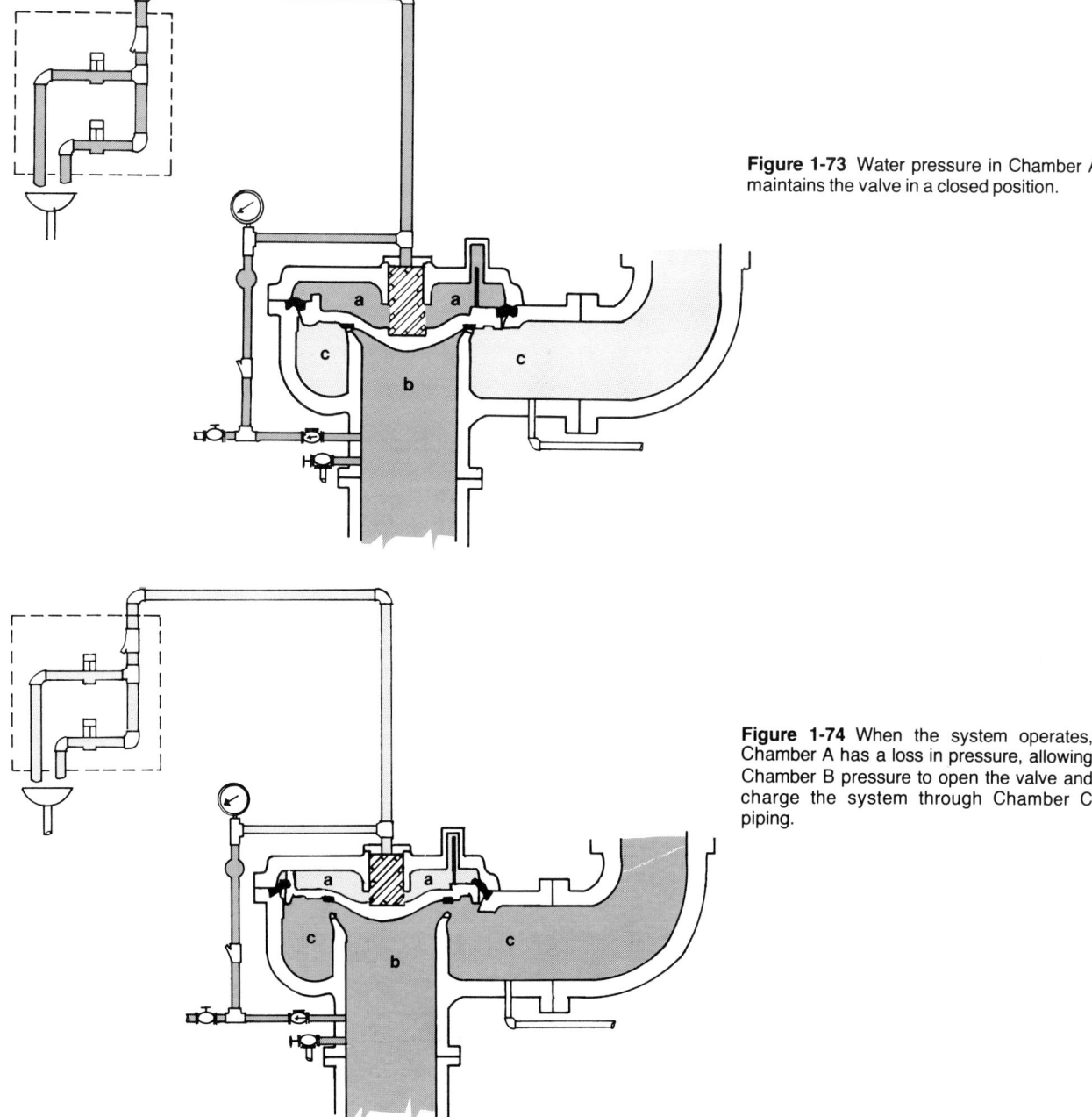

Figure 1-73 Water pressure in Chamber A maintains the valve in a closed position.

Figure 1-74 When the system operates, Chamber A has a loss in pressure, allowing Chamber B pressure to open the valve and charge the system through Chamber C piping.

Inspecting Deluge or Pre-Action Systems

Deluge and pre-action system inspections follow the same guidelines as those performed on wet and dry systems, in regard to piping and valves. During normal conditions the inspector should insure that:

- The main supply control valve is open and supervised in the open position.
- Any valves in lines to alarm devices are wide open.
- The alarm bypass valve is closed.
- The pressure gauge valves are open.
- Pressure gauges, water and air (if used) are indicating adequate pressure.
- Check the drip valve as described under "Dry-Pipe Sprinkler System" for those valves so equipped.
- No air leaks are evident in pneumatic detection systems.
- Fire detection devices are not damaged or corroded.
- Air maintenance equipment is working properly.
- There are no air leaks in pre-action system piping.
- There are no water leaks from preprimed deluge system sprinkler plugs or caps (Figure 1-75).
- Piping and hangers are in good condition, free of corrosion and not subject to physical damage.
- Sprinkler head cabinets contain a sprinkler wrench and the appropriate number and type of replacement heads.

MAIN DRAIN AND ALARM TESTS

Step 1: Record static pressure.

Step 2: Open 2-inch main drain slowly to avoid tripping the valve (Figure 1-76).

Figure 1-75 During inspections of preprimed deluge systems, check for leaks from sprinkler caps or plugs.

Figure 1-76 Open the main drain slowly to avoid tripping the valve during a main drain test.

Step 3: Record the waterflow pressure.

Step 4: Close main drain slowly (Figure 1-77).

Step 5: Open the alarm bypass valve. When the alarms sounds return it to the closed position (Figure 1-78).

TRIP TEST

Step 1: On a small deluge system replace open sprinklers with standard sprinkler plugs or caps (Figure 1-79). On large systems where the heads cannot be replaced or plugged, close the main water supply (OS&Y) to within two turns from closed, trip the valve and immediately close the valve.

Step 2: Activate the system by using a heating device on a detector or by manual means (Figure 1-80).

Figure 1-78 Test the alarm by opening the alarm bypass valve until the alarm operates.

Figure 1-77 After recording the flow pressure, close the main drain.

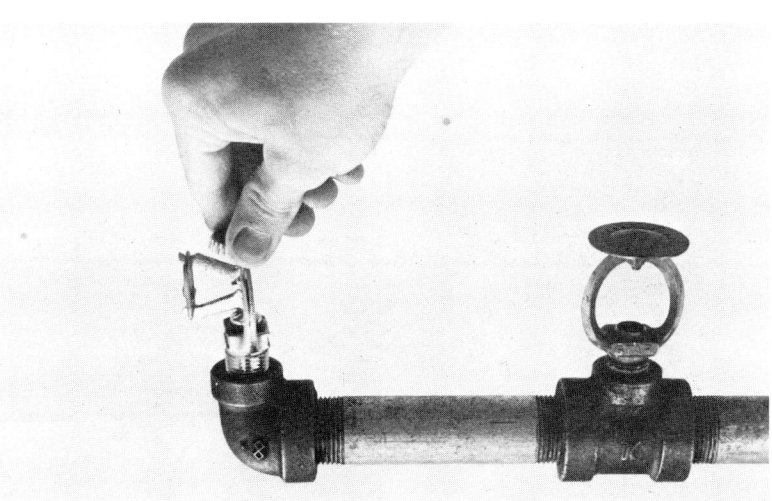

Figure 1-79 Replace the open sprinklers on a small deluge system before running a trip test.

Figure 1-80 Operate the manual pull to activate the valve.

54 PRIVATE FIRE PROTECTION

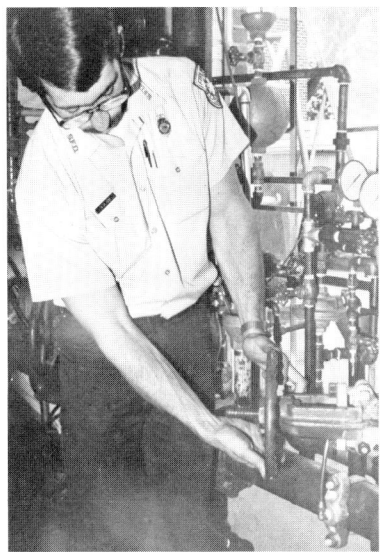

Figure 1-81 After closing the main water supply, use the main drain valve to drain the system.

Step 3: Close main water supply (OS&Y) valve and drain the system using the 2-inch main drain valve (Figure 1-81).

Step 4: Open valve faceplate, clean the valve seat and any debris from valve housing (Figure 1-82).

Step 5: Unlatch the clapper and reset the valve* (Figure 1-83).

Step 6: Reinstall the faceplate* (Figure 1-84).

Step 7: On pre-action systems, if the pipe integrity is supervised, pressurize system with air to the proper level.

Step 8: Restore water pressure by opening the supply control valve with the 2-inch main drain open to avoid tripping the valve (Figure 1-85). When the supply control valve is completely open, slowly close the 2-inch main drain. Check gauges.

Step 9: On deluge systems, remove plugs or caps from sprinkler heads or reinstall open heads.

*Some deluge and pre-action valves do not require resetting by removal of the faceplate. Refer to manufacturer's instructions for proper resetting procedure.

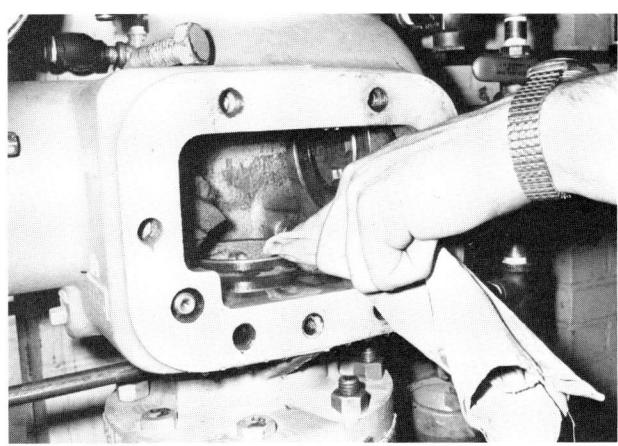

Figure 1-82 Remove faceplate cover, clean the valve seat and the debris from the valve housing.

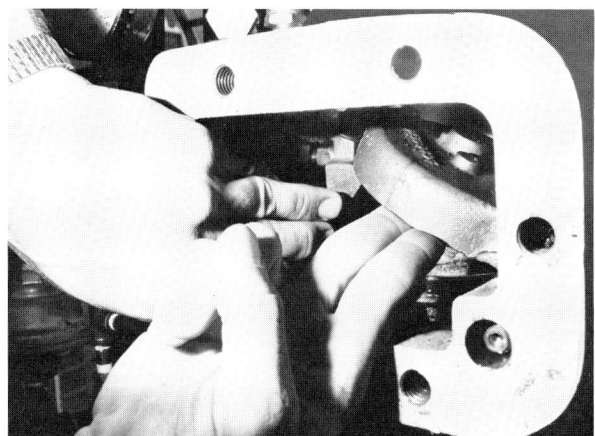

Figure 1-83 Reset the valve by unlatching the clapper.

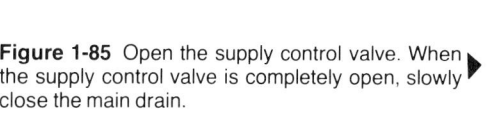

Figure 1-84 Replace the valve cover.

Figure 1-85 Open the supply control valve. When the supply control valve is completely open, slowly close the main drain.

Frequency of Inspecting and Testing of Deluge and Pre-Action Systems

Deluge and pre-action systems should be visually inspected at least monthly; however, since visual inspections provide continued familiarization and early problem discovery, a weekly inspection is recommended.

The main drain test should be performed at least annually although a quarterly test is suggested.

Inspect monthly — test at least annually

The trip test should be accomplished at least annually. The inspector should visually inspect the system during each drain and alarm test. A main drain test and alarm test should be performed in conjunction with each trip test. A written record of all tests and inspections should be made, compared to earliest tests and filed.

SPRINKLER SYSTEM ALARM ATTACHMENTS

Signaling attachments used for indicating waterflow in a sprinkler system may be responsive to the movement of water through the system or variations in pressure.

The point of connection for the pressure-operated type of devices will depend on the type of system. For a wet-pipe system, the pressure switch is connected directly to the alarm line from the alarm check valve (Figure 1-86) or may be affixed to a retard chamber which is fed through the alarm line. The switching device may incorporate its own retard feature which will eliminate the necessity of an extra retard chamber. On a dry system, a pressure switch may be attached to the system above the valve clapper mechanism.

Figure 1-86 Alarm line pressure switch for electrically signaling water flow.

The movement of water through a wet system may also be sensed with a paddle or flow vane type of switch. This type of device may also include a retard feature, but should not be used on a dry or deluge system in that the vane that is inserted in the pipe may be sheared off by the force of the water when the system activates. A relatively new type of device that detects flow, but does not penetrate the pipe to which it is affixed may be used on either wet or dry systems. This is a sonic (sound) system.

56 PRIVATE FIRE PROTECTION

◀ Figure 1-87 Signal initiating device.

Figure 1-88 Coded sprinkler alarm ▶ device.

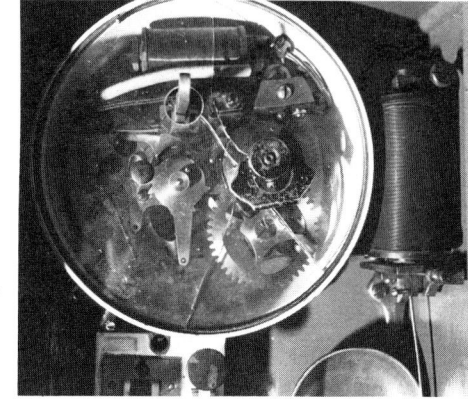

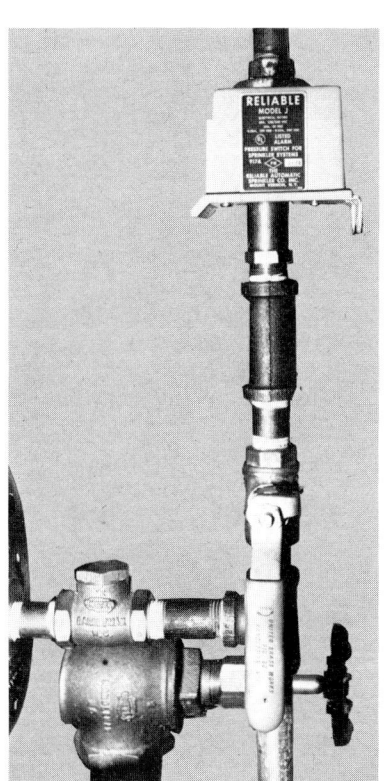

Figure 1-89 Non-coded sprinkler alarm device.

The signal-initiating device may be a simple switching mechanism in an alarm current (Figure 1-87), or may be an integral part of a coding device suitable for direct connections to a McCulloh signaling coder. The signal produced is treated as a fire warning.

Sprinkler system alarm attachments send either coded or noncoded signal impulses indicating waterflow. Certain attachments may provide automatic supervision that sends a signal impulse when a sprinkler function becomes inoperative or when a change in the system affects the water supply or distribution. A coded sprinkler attachment is shown in Figure 1-88 and a noncoded attachment is shown in Figure 1-89.

Supervisory attachments monitor the sprinkler system for operating condition. The most common supervising services are as follows:

- When the main water control valve is being closed, the attachment operates a signal mechanism.
- Air and water pressure attachments are used to indicate any major change in pressure.
- A thermostatic attachment which responds to low temperature conditions in water supply or building interior.

Careful inspection, service testing and maintenance are essential if a supervisory system is to perform reliably. To insure that proper remedial action is taken without delay, the system inspectors and operators at fire stations and other points where signals are transmitted, received and recorded must be able to interpret the supervisory signals coming in (such as a low air pressure, waterflow or the closing of an outside screw and yoke control valve).

RESIDENTIAL SPRINKLER SYSTEMS

One of the innovations in the field of private fire protection, primarily in apartment towers, has been the installation of residential sprinkler systems. These systems are designed and constructed essentially the same as most other sprinklers, with the exception of two major changes: first, the piping of these systems, instead of being steel pipe, may be copper or, in some jurisdictions, even CPVC. Second, residential sprinkler systems generally do not have a fire department connection and are designed to function with a limited number of heads activated.

Most of the departments that have adopted residential sprinkler system ordinances have used NFPA Standard No. 13D, or a modification of it or Standard 13. The basic reason for most communities adopting this ordinance is to increase private fire protection so that public fire protection expenditures can be held to a minimum. Tests have revealed that the atmosphere in an area with residential sprinklers remains safe and will support life as the fire initiates and burns, and after extinguishment. Communities that have adopted these ordinances most generally are ones that are in the master planning process or are facing severe financial difficulties.

Residential sprinklers maintain safe life supporting atmospheres

Requiring residential sprinkler systems generates certain activities that many fire departments are not currently involved in. Inasmuch as these systems have to be carefully calculated, they often require considerably more activities by the Fire Prevention Bureaus in the plan-checking process. In addition, they create a new need for fire prevention personnel to inspect installations in single-family dwellings. Temporarily, this shifts the focus on man hours from fire suppression to fire prevention. While policy varies from locale to locale, many communities also devote more time to an annual reinspection of such systems to insure their reliability and operational effectiveness when needed.

From a tactical point of view, the residential sprinkler system does not normally generate any new need other than the fact that fires can be combated with less manpower and small fire streams. This is because fires generally never reach the flashover state or total structural involvement in occupancies with such systems. There may be an increased need for the use of salvage appliances because of the obvious problem of exposure to additional water damage.

The fire department will need to develop a public relations effort to encourage residential sprinklers and the adoption of a code requiring them. A number of arguments must be addressed, including cost, unsightly appearance detracting from the aesthetics of the home, and the fear of premature discharge or excessive water damage. Advantages above the obvious life-safety value and loss reductions in case of fire include reduced insurance costs and higher property evaluations. The fire department will also have to recommend procedures for maintenance, testing, and restoration of residential sprinkler systems.

Develop public education programs to encourage residential sprinklers

SPECIAL EXTINGUISHING SYSTEMS

- Carbon Dioxide Systems
- Halogenated Agent Systems
- Dry and Wet Chemical Systems
- Foam Systems

CARBON DIOXIDE EXTINGUISHING SYSTEMS

Carbon dioxide (CO_2) is one of the most common chemical compounds found in nature. Carbon dioxide is produced by combustion of carbonaceous materials, fermentation, synthesis of ammonia and other chemical processes. It is also found in relatively pure form in certain gas wells. It is available world-wide and exists as a gas, as a liquid under pressure and in solid form as dry ice. Carbon dioxide is a very effective extinguishing agent.

When liquid carbon dioxide is discharged from a container, it immediately flashes to a mixture of vapor and fine dry ice particles. When the discharge velocity is controlled by a suitable nozzle shell, the discharge stream can be effectively used to extinguish a liquid surface fire. It is easy to see that the discharge cloud from a portable extinguisher separates the flames from the fuel long enough to provide complete extinguishment. The dry ice particles in the discharge increase the density of the stream. It is clear, however, that the fire must be completely extinguished during the operation or it will flash back after the discharge is stopped because the cloud of gas is quickly dissipated.

Carbon dioxide extinguishes by oxygen dilution

Fuel, heat and oxygen must all be present to start and sustain combustion. Since the fuel cannot be removed and the cooling effect of carbon dioxide is minimal, extinguishment by application of carbon dioxide is based upon oxygen dilution. In order for a fire to continue, combustion at the flame front must produce enough heat to bring the incoming fuel, the oxygen and any inert gas present up to the ignition temperature. When more than 29 percent carbon dioxide is added to the atmosphere, the heat generated is not sufficient to produce this result.

In any gaseous or surface fire, heat is also lost rapidly by radiation, and, therefore, the flame suddenly goes out when the critical concentration of inert gas is achieved. The holding or cooling time for liquid fuel fires is generally very short because the liquid fuel itself normally cannot be heated to a point approaching its auto-ignition temperature, but instead is limited to its boiling temperature by the cooling effect of evaporation. Consideration must, however, be given to the cooling time requirements for other materials, such as metal parts, that may be heated above the ignition temperature by the fire and can cause reignition.

There are a few exceptions where liquid fuels are capable of being heated above the auto-ignition temperature. Oils used in deep fat fryers, for example, have boiling temperatures in excess of 700°F. This is well above the auto-ignition temperature which may be as low as 600°F. Such fires are notoriously difficult to extinguish because of the long ambient cooling time required.

A different type of fire extinguishing problem is presented by ordinary combustibles such as wood, cloth and paper. These materials present hazards where there is no temperature-limiting effect. Deep-seated fire may develop within thermally insulated areas where very little heat can be dissipated by direct radiation. The concentrations and cooling time for extinguishing these types of materials depend upon both the nature of the material and the mass of material present as related to the time required for heat dissipation. Thus, carbon dioxide is not very effective as an extinguishing agent for Class A materials.

Major advantages of a carbon dioxide system are:

- Effective for extinguishment of open and closed tanks of flammable liquids.
- Very effective for total flooding of areas containing electrical hazards or flammable liquids.
- Halts active combustion of Class A fires.
- Non-conductor of electricity.
- Does not damage or leave a residue on high-value contents such as records, furs or electrical machinery.

Some disadvantages to a carbon dioxide system are:

- After the carbon dioxide dissipates, reignition can occur.
- Generally, it will not extinguish a smoldering Class A fire.
- Not economical for large special hazard areas.
- Accomplishes little cooling, only about one-tenth as effective as water.
- Reduction in oxygen content can cause asphyxiation.
- Discharge noise and limited visibility can cause panic.

Normally, with installed systems carbon dioxide is stored outside the hazard area in cylinders. Each cylinder is equipped with a valve, siphon tube assembly and a discharge head. The discharge heads are connected to a manifold by flexible connectors (braided metal hose). The manifold connects to the discharge piping and, if both a main and reserve supply are provided, check valves are used to isolate one from the other. Each bank of cylinders containing three or more cylinders is equipped with two "pilot" cylinders which are released by the system controls. The remaining cylinders are "slave" cylinders in that their discharge heads are operated by pressure in the manifold. When the pilot cylinders are released, they discharge into the manifold and the pressure releases the balance of the cylinders. Figure 2-1 shows the piping arrangement and cylinders of the combination pilot and slave cylinder carbon dioxide system.

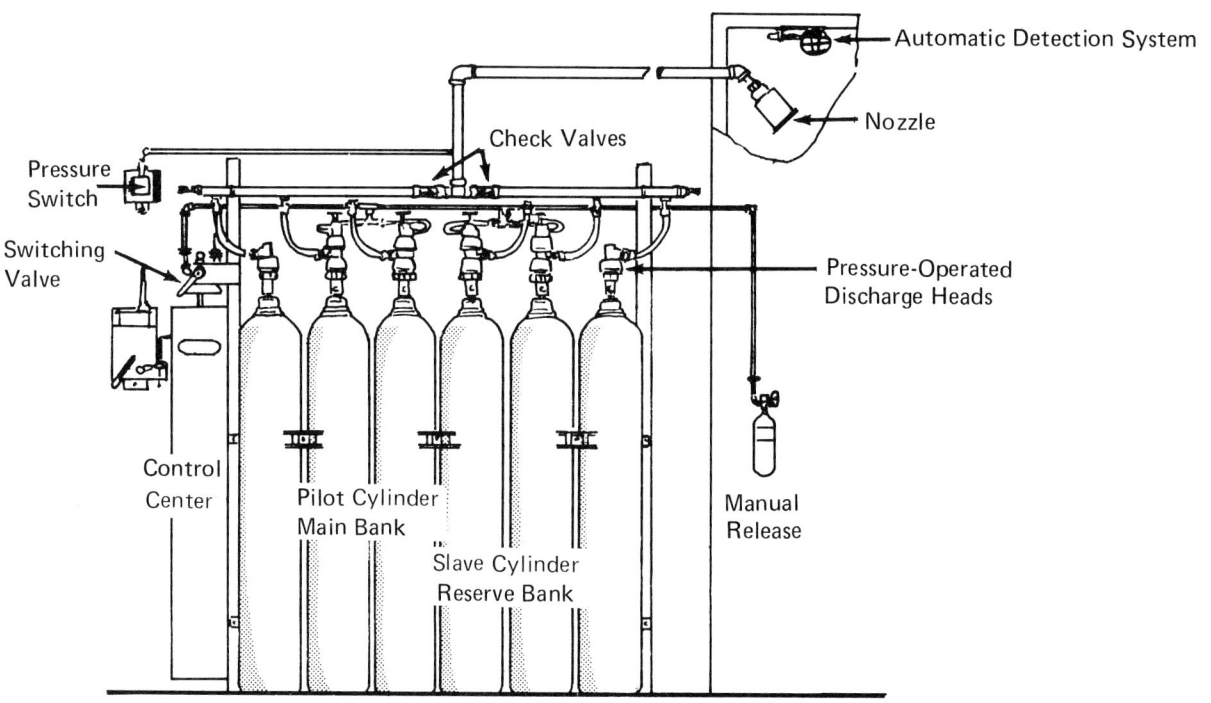

CO_2 STORAGE AND SUPPLY SYSTEM

Figure 2-1 A high-pressure carbon dioxide storage bank consists of a set of controls and switches, main and reserve supply cylinders and a manual operation station. *(Courtesy of Cardox Div. of Chemtron Corp.)*

A discharge delay to provide a predischarge evacuation alarm for personnel and to permit the shutdown of ventilation fans is accomplished in several ways depending upon the equipment used. This warning alarm is important since a carbon dioxide laden atmosphere will not support life. There is some slight variation in equipment between manufacturers. It can be done with electric actuation, using a time delay relay. Otherwise, carbon dioxide pressure is used through a pneumatic timer to operate a blocking valve in the manifold or discharge piping. The blocking valve is opened by the timer after the predischarged period. This releases the carbon dioxide into the discharge piping. A pressure switch for the carbon dioxide alarm which is operated by pressure ahead of the blocking valve accomplishes the predischarge alarm or equipment shutdown. One manufacturer, using a separate pressure source for actuation, places the pneumatic timer between this pressure source and the pilot cylinder discharge heads to delay actuation. With the carbon dioxide system, additional controls are necessary for operation.

In electrically operated systems, a solenoid or an electrically initiated squib is energized by the detection system and operates the pilot discharge valves. These devices can be mounted directly on the pilot discharge valves or in a separate control cabinet. Electrically operated systems permit the use of any one or a combination of the numerous types of fire detection devices and systems such as thermostats, rate-of-rise detectors, products-of-combustion detectors and smoke detectors. When fire occurs, the detectors cause the system release to operate.

Electrically operated systems are released by detectors

A pneumatic rate-of-temperature-rise system is an alternative means of automatic operation where it is desirable to function independently of external power. Basically, the system consists of heat actuators connected by small-bore tubing to pneumatic discharge releases. The heat actuator is a heat-sensitive air chamber. A rise in temperature results in an increase in pressure within the actuator. If the temperature increases at a normal rate, the pressure is bled off through a vent in the pneumatic release. In the event of a fire in the area protected, however, the temperature rise becomes excessive, the vent cannot bleed off the pressure fast enough and the pneumatic release operates to discharge the system.

The pneumatic rate-of-temperature-rise system is completely self-contained and does not depend upon any outside source of energy for actuation. It is considered a fast means of detection for a flash fire and is usually recommended for such applications. There are some exceptions where hazards have abnormal heat conditions present, such as oil quench tanks, kitchen range hooks and ducts, and ovens, where it is more suitable to used a fixed-temperature means of fire detection.

The fusible-head (pneumatic) automatic system consists of fusible heads located in the hazard area and connected by gas-tight tubing to a release valve. In the event of fire, the heads melt at a predetermined temperature and release the pilot pressure, causing the system to discharge.

In a fusible link system, automatic actuation is produced by fusible link detection. Fusible link detectors are connected by cable run in conduit and corner pulleys to a spring-loaded release device at the supply unit. These detectors are mounted in hazardous areas. If the temperature in the hazard area rises above the predetermined level, the link melts, releasing the cable, which initiates discharge of the system (Figure 2-2).

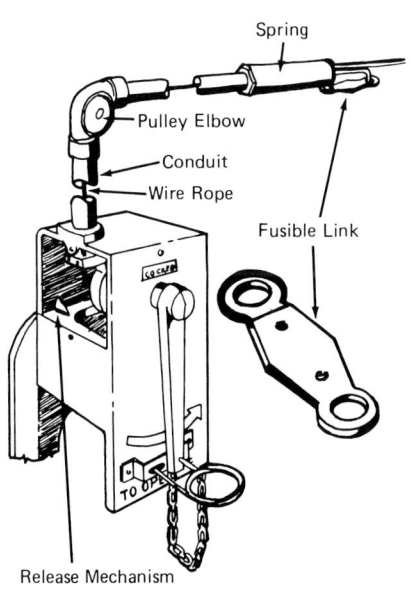

Figure 2-2 Carbon dioxide systems can be released mechanically, using fusible elements.

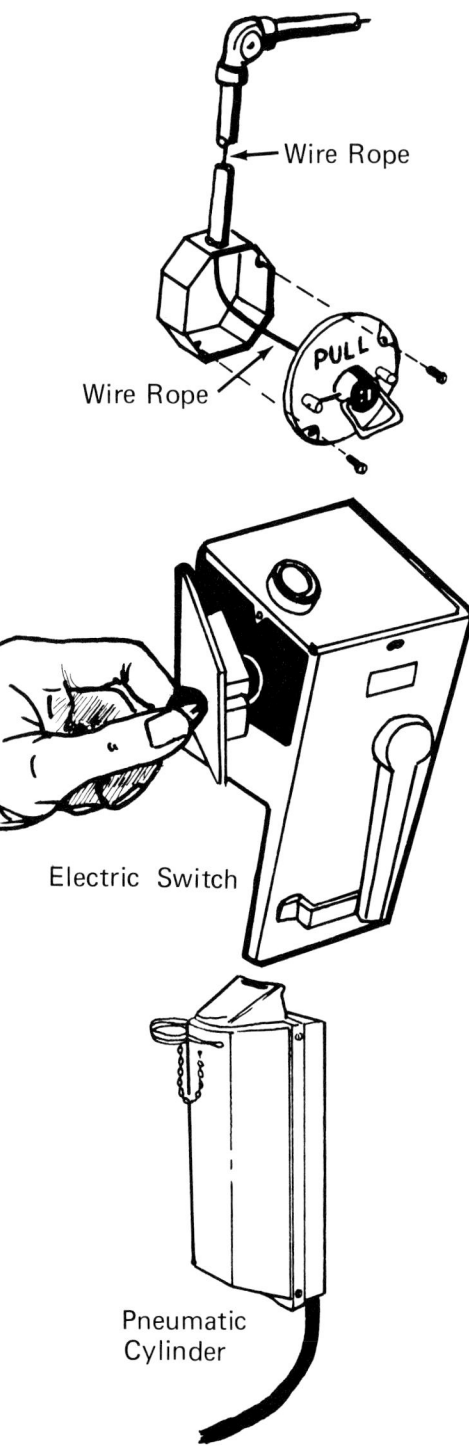

Figure 2-3 A variety of manual actuators are available to initiate system operation from remote locations. These include cables, electric switches, or pneumatic cylinders connected to the storage control.

Some systems have a remote manual actuator which operates by pull cable, electric switch or pneumatic cylinder. The remote pull cable is connected in conduit and on pulleys to an actuator on the pilot discharge heads. It offers total manual system activation under possible periods of power outage or direct control in the event of the failure of an electrical component (Figure 2-3).

Another mechanical actuator is the pneumatic cylinder which discharges gas pressure through piping directly to the pilot discharge valves. These devices also have the feature of operating when power outages or electrical failures occur.

The third remote actuator device is the electric switch or push button station. The switch triggers an electrically operated pilot discharge valve. These devices are commonly wired in parallel with the detection system. They should be supervised circuits with an approved auxiliary power source.

The systems have the added feature of direct manual release located on the pilot discharge valves and selector valve when used. These are mechanically operated devices that totally override automatic or remote actuators. These should be accessible and properly placarded.

When more than one hazard is protected from a central storage location, selector or direction valves are used. These valves have direct manual actuators and are operated by the automatic and remote devices basically in the same manner as the cylinder pilot discharge valves are operated.

These fixed systems employ two types of storage of carbon dioxide, Low-pressure storage consists of refrigerated tanks holding liquid carbon dioxide at 300 psi at 0°F. The storage tanks range in size from 500 pounds to 125 tons (Figure 2-4). Low-pressure carbon dioxide storage containers will be manufactured, tested and marked in accordance with current specifications of the American Society of Mechanical Engineers (ASME) Code for Unfired Pressure Vessels. The design working pressure will be at least 325 psi. In high-pressure storage, liquid carbon dioxide is stored in 50- to 100-pound metal cylinders (Figure 2-5). The number of cylinders varies with the size of the area being protected and the method of carbon dioxide application. These cylinders contain carbon dioxide at 850 psi at 70°F. Either of the two may be used as shown in Figure 2-6. High-pressure cylinders must meet Department of Transportation specifications for test (DOT-3A, 3AA-1800 or higher). They will not be recharged if the last hydrostatic test is over five years old.

The entire layout of the fixed carbon dioxide system is very similar to the dry or deluge sprinkler system. In both the high- and low-pressure carbon dioxide systems the storage units are

Special Extinguishing Systems

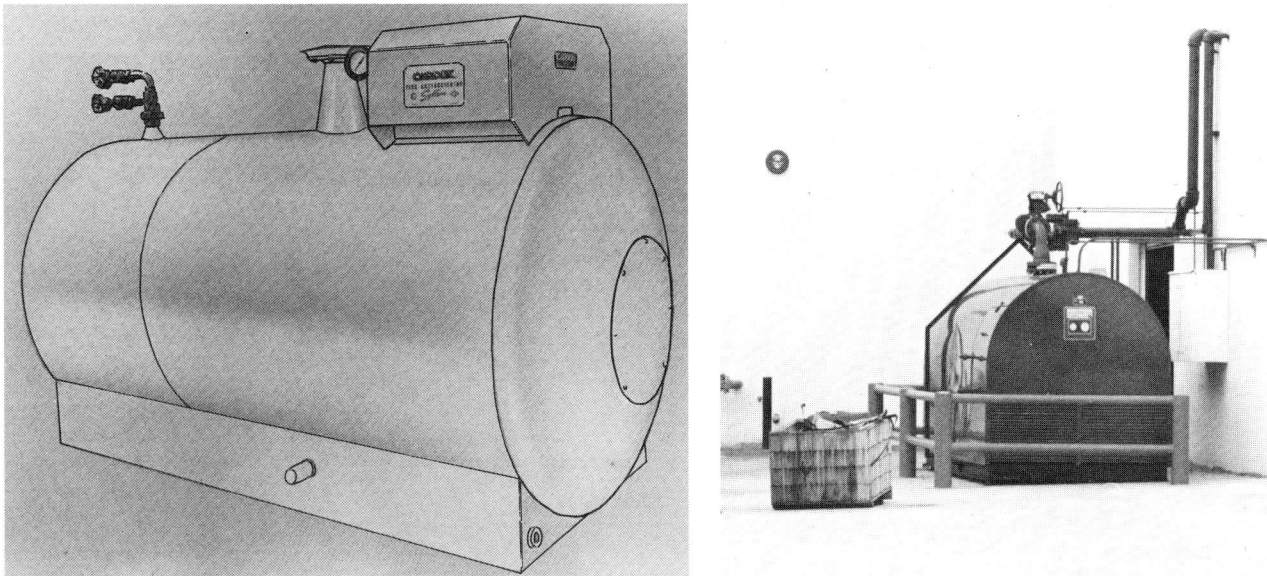

Figure 2-4 Low-pressure storage tanks for carbon dioxide systems are of steel or fiberglass construction. These refrigerated tanks contain liquid carbon dioxide at 0°F and 300 psi. *(Courtesy of Cardox Division of Chemtrom Corp.)*

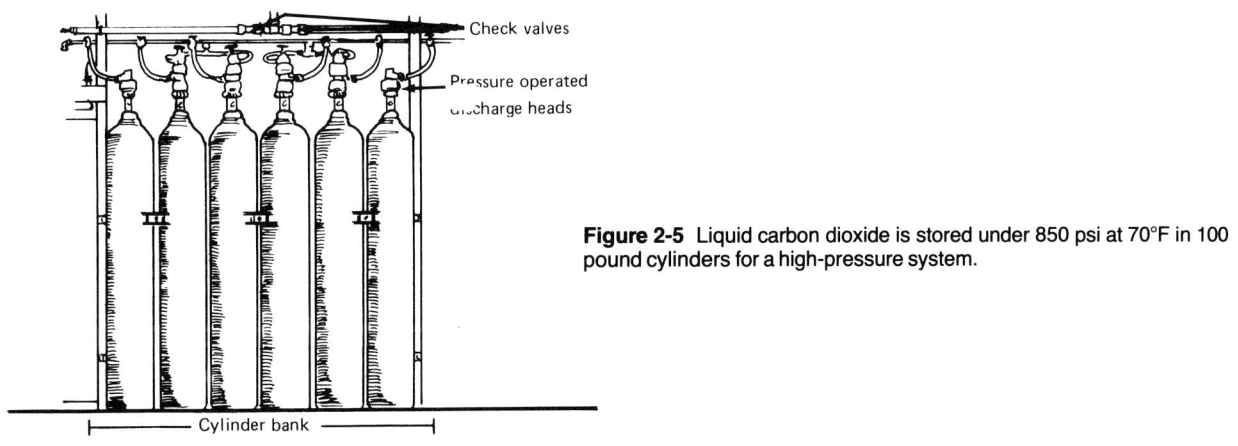

Figure 2-5 Liquid carbon dioxide is stored under 850 psi at 70°F in 100 pound cylinders for a high-pressure system.

CO₂ STORAGE AND SUPPLY SYSTEM

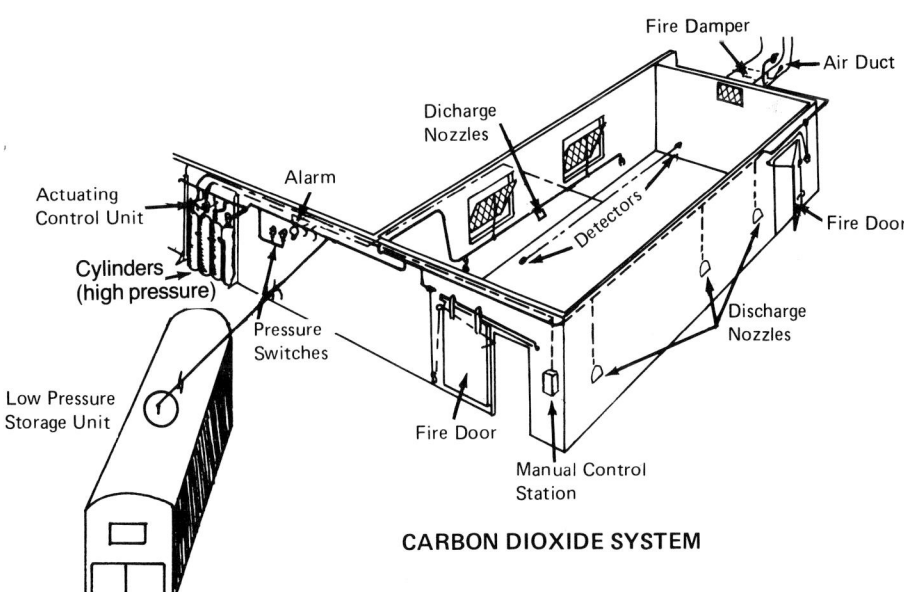

Figure 2-6 A system may be installed that includes both low- and high-pressure storage of carbon dioxide. *(Courtesy of Walter Kidde & Co., Inc.)*

CARBON DIOXIDE SYSTEM

piped into discharge orifices or nozzles. Both systems may be operated either automatically or manually. Automatic operation is normally controlled by quick-operation detection devices. Automatically operated systems must also have an independent means for manual operation. Supervision for carbon dioxide systems is recommended. This should include a signal to indicate any system failure and an audible or visual signal to indicate the system is operating and that the area should be evacuated.

Four methods used in the application of carbon dioxide with fixed systems are briefly described as follows:

1. Total Flooding. This consists of replacing the entire atmosphere of an enclosure with carbon dioxide (Figure 2-7).

Figure 2-7 This large quench tank is protected by a total flooding low-pressure carbon dioxide system. *(Courtesy of Cardox Division of Chemtron Corp.)*

2. Local Application. In this method, the fire is extinguished by discharging carbon dioxide directly into the fire or on the burning material.
3. Extended Discharge. This method of application floods the area initially with carbon dioxide and then releases additional amounts to maintain the desired level of concentration.
4. Hand Hoselines. Handlines are used manually as first-aid extinguishing equipment within a limited area. These handlines may also be used to apply carbon dioxide in any of the other above methods.
5. Standpipe and Mobile Supply. This is a mobile supply of carbon dioxide which can be quickly moved into position and connected to a piping system to supply fixed nozzles or hoselines for local application or total flooding.

The extinguishment of fires using hand hoselines or mobile supplies depends upon ability and application techniques of the individual operator. All personnel who will use this equipment during a fire must be trained to properly operate it and in the fire fighting techniques of applying carbon dioxide.

FIRE DEPARTMENT OPERATIONS

A major concern with any carbon dioxide system is the possibility of asphyxiation to persons occupying an enclosed area after discharge. This danger is especially important to firefighters since it is assumed that the system will have already discharged by the time the fire department arrives. Each firefighter entering the area should be protected with self-contained breathing apparatus. The area should be searched immediately for persons who were not able to evacuate the area. Fire fighting may now begin, based on the officer's evaluation of the scene. If fire still exists, it may have to be attacked with another extinguishing agent. However, if the fire has been extinguished, it may be desirable to seal the area and allow the high concentration of carbon dioxide to remain. This will be the decision of the fire officer since heavy smoke may require immediate ventilation to reduce damage.

Wear self-contained breathing apparatus when entering the CO_2 flooded area

INSPECTING AND TESTING OF THE SYSTEM

Carbon dioxide systems should be installed in accordance with NFPA Standard No. 12, *Carbon Dioxide Extinguishing Systems*. Inspect and test to insure that all enclosures designed to confine the gas are maintained in good condition, and that all automatic shutters used in connection with enclosures are fully operative. See that all piping and apparatus located at the carbon dioxide storage units are adequately protected against accidental damage and obstructions. Carbon dioxide storage units located outdoors should be protected from inclement weather by non-combustible enclosures. Check any changes that may have been made in the hazard, surroundings and apparatus that affect the degree of protection required.

Check the amount of gas in storage cylinders by weighing them in accordance with schedules in applicable codes. In the high-pressure systems, replace or recharge any cylinder showing a loss of gas of more than ten percent by weight of the original content. Refill any tank in the low-pressure system showing more than ten percent loss of contents if its capacity is equal only to the amount of gas required to protect the largest hazard. For systems designed to give immediate re-protection for the largest hazard after a fire is extinguished, the quantity of gas to be maintained should be equal to twice that required for the largest hazard. Test automatic and manual release devices for good operating condition, in accordance with the testing procedures outlined for that particular system. Frequent visual checks of the system should be made.

Refill any tank that is down ten percent

Due to the sophisticated nature of carbon dioxide systems, expert consultation should be sought concerning questions of design and operation.

HALOGENATED AGENT EXTINGUISHING SYSTEMS

Halon is a generic term for halogenated hydrocarbons and is a chemical compound that contains carbon plus one or more elements from the halogen series (Fluorine, Chlorine, Bromine or Iodine). While a very large number of halogenated compounds exist, only a few are used to a significant extent as fire extinguishing agents. Halogenated vapor is a nonconductor of electricity and is effective in fighting surface fires such as flammable liquids, most solid combustible materials and electrical fires.

Halogenated agents work chemically to extinguish fire. They stop the combustion process itself by breaking the chain reactions, preventing further fire propagation. This chemical fire-stopping action happens with only a low concentration of halogenated agent required to extinguish most fires. Application of the agent may be applied locally by using a compressed bottle of noncombustible gas similar to a carbon dioxide fire extinguisher. This type of application is effective in controlling or extinguishing surface fires involving flammable liquids, solids, or gases, such as dip tanks, quench tanks, spray booths, oil-filled transformers or vapor vents.

Total flooding applications require a fixed system which expels the agent into an enclosed atmosphere (Figure 2-8). The enclosed area must be sealed in such a manner as to maintain a

Figure 2-8 A closed vault for record protection is an ideal application of a halogenated system.

certain concentration of the agent over a period of time. Total flooding systems are used to extinguish 1) surface fires involving flammable liquids or solids, 2) flammable vapor-air mixtures and 3) deep-seated Class A fires.

Halon 1301, Bromotrifluoromethane and Halon 1211, Bromochlorodifluoromethane, are the most common halogenated agents and are listed by NFPA Standard Nos. 12A and 12B for use in systems. Halogenated extinguishants may be used in concentrations up to ten percent where people are normally present, but Halon 1211 may not be used in areas normally occupied by people. Since it is a vapor and not a spray fog, or snow, concentrations of less than ten percent of Halon 1301 permit sufficient time for personnel to exit from the hazard area without harm. Halon 1211 is applicable to areas only occasionally visted by persons, such as switch houses, pump rooms and vaults due to the health hazard.

Halon 1301 is a particularly effective extinguishing agent for areas containing electronic equipment such as computer enclosures. Fires in such hazard areas can be quickly extinguished with a minimum of fire damage. The halogenated agent itself leaves no residue and does not cause electrical short circuits or cause damaging corrosion of the equipment. Halon 1301 has been approved for use in normally occupied areas such as schools, museums and hospitals. Both Halon 1301 and Halon 1211 have found applications in laboratories, petroleum refineries, power-generating stations, ships, airplanes and chemical plants. A typical Halon 1301 total flooding system plus a schematic drawing showing essential components is shown in Figure 2-9.

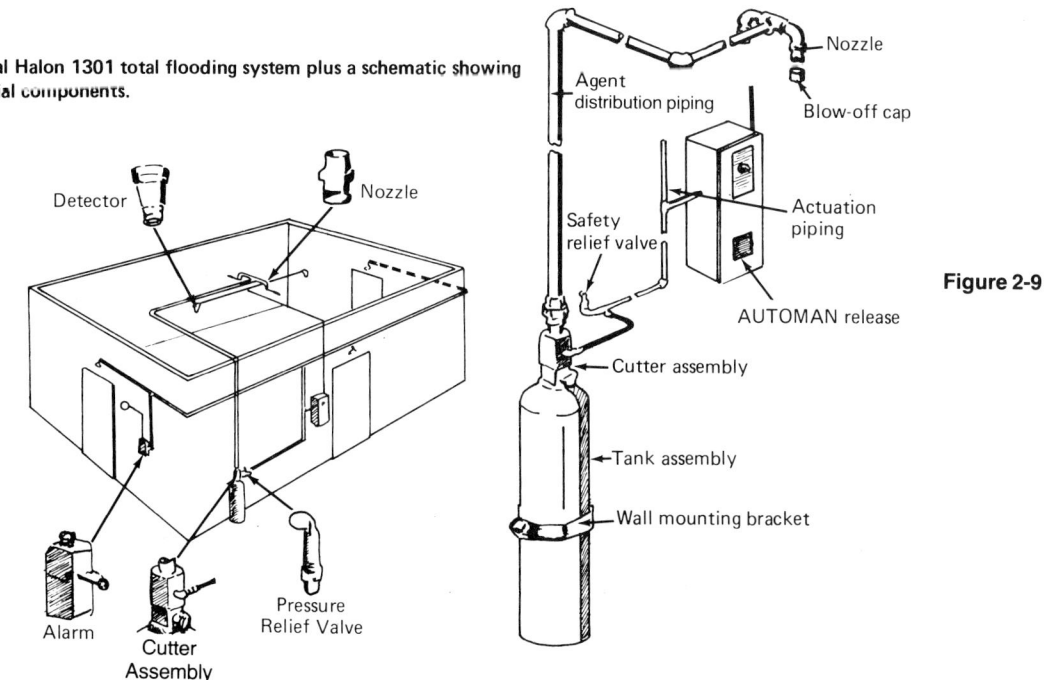

Typical Halon 1301 total flooding system plus a schematic showing essential components.

Figure 2-9

"CLEAN AGENT" FIRE EXTINGUISHING

SYSTEM DESCRIPTION

Halogenated agents are used in both total flooding and local application systems. These systems require automatic detection and actuation (Figure 2-10). Agents are stored in containers as liquified compressed gases which are pressurized with nitrogen to promote rapid discharge over wide temperature ranges.

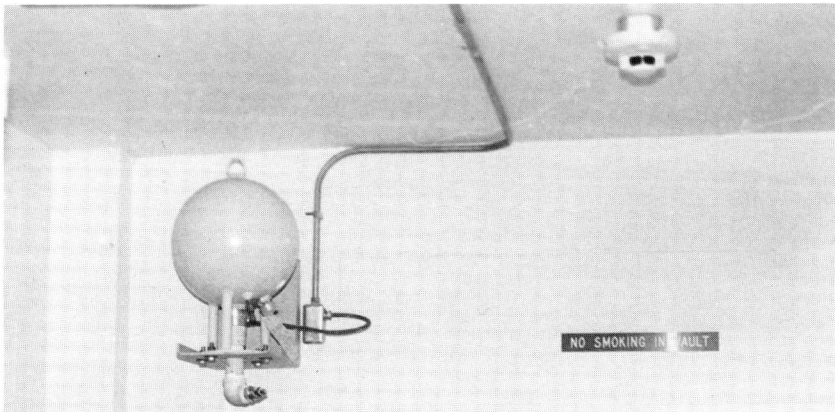

Figure 2-10 A typical halogenated system has automatic detection equipment and a liquefied halon storage container.

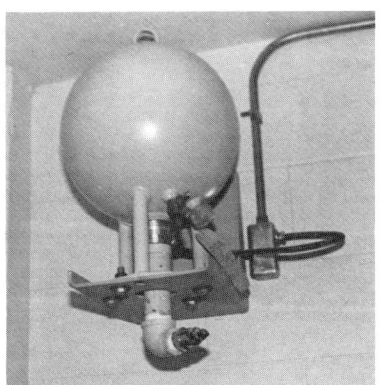

Components common to all halon systems are containers, actuators, (Figure 2-11) nozzles, detectors, manual releases and control panels. Selector valves are sometimes used for protection of multiple hazards. The shape of containers can vary. Tank capacities range from 5 to 250 pounds, with larger sizes available for special applications. The amount of agent in a container and the number of containers will depend on the system design. Containers are designed to meet the requirements of the U.S. Department of Transportation (DOT) and the Canadian Board of Transportation Commissioners. Pressure gauges are frequently used to indicate the pressure in the container (Figure 2-12).

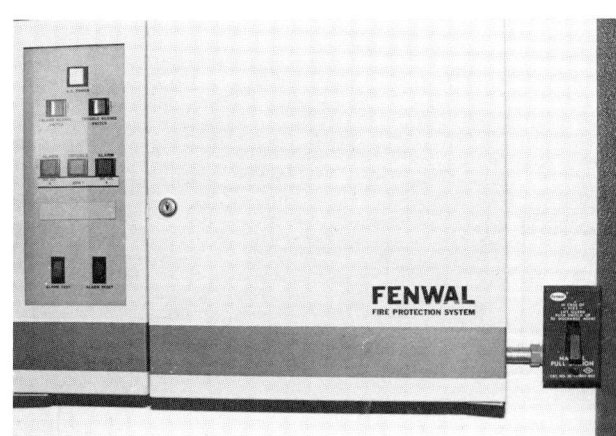

Figure 2-11 The halogenated nozzle is somewhat different than those used by other systems. Manual controls for activating the system are also installed.

Figure 2-12 Container pressure is measured by cylinder or tank pressure gauges.

All containers have one or more valves to permit release of the agent into the hazard area or distribution system. Valves may be actuated mechanically, by liquid or gas pressure or electrically. Valves used in halogenated agent systems are larger in diameter than those used with other agents due to the higher discharge rates required for halogenated agents. Discharge nozzles are chosen from those listed for use with these agents according to the design requirements of the system. The discharge nozzle consists of the orifice, which determines discharge rate, and any associated horn, shield or baffle.

FIRE DEPARTMENT OPERATIONS WITH HALON

Underwriters Laboratories' classification of comparative life hazard of various chemicals rates Halon 1301 as less toxic than carbon dioxide. The NFPA Standard No. 12A, *Halogenated Extinguishing Systems of Halon 1301*, felt that a seven percent concentration was safe for normally occupied areas without restrictions. Greater concentrations will require evacuation and other considerations. Tests have shown that concentrations of 20 percent can be hazardous. With these facts, there appears to be no need for concern regarding halogenated systems; however, products of decomposition after discharge of the agent present a more serious hazard. Tests have discovered that decomposition products such as hydrogen fluoride, hydrogen bromide and other carbonyl halides are toxic. For this reason it would be best to require positive-pressure, self-contained breathing apparatus before entering an area protected by halogenated agents. Once entry has been gained, a search should always be initiated to remove personnel from the immediate and adjacent areas. Operations will be similar to that with carbon dioxide total flooding

Wear self-contained breathing apparatus in halogenated fire areas

systems already discussed. Ventilation may be necessary to clear the area of remaining halon vapors and decomposition products before permission is given to re-enter the area.

INSPECTING THE SYSTEM

Procedures for inspecting a halogenated agent system should basically follow those of the carbon dioxide fixed system. Differences in inspection procedures should be established by consulting the factory representative for the system. Once the system is discharged, it should be placed back into service by a factory representative.

Due to the sophisticated nature of halogenated agent protection systems, expert assistance may be required to check and approve the adequacy of design of the systems. Halogenated agent systems are not effective on all types of fires. For example, they are ineffective on fuels containing their own oxidizing agent. These fuels, such as sodium or potassium, are reactive metals and, as with metal hydrides, they are very reactive to halogenated agents.

Figure 2-13

SAMPLE HALON NUMBERS FOR COMMON HALOGENATED AGENTS

Halon 1011	CH_2BrCl	Chlorobromomethane (LB)
Halon 1202	CBr_2F_2	Dibromodifluoromethane
Halon 1211	$CBrClF_2$	Bromochlorodifluoromethane (BCF)
Halon 1301	$CBrF_3$	Bromotrifluoromethane
Halon 2402	$CBrF_2CBrF_2$	1,2-Dibromotetrafluoroethane

HALON NUMBERING SYSTEM

1st Digit	No. of Carbon Atoms
2nd Digit	No. of Fluorine Atoms
3rd Digit	No. of Chlorine Atoms
4th Digit	No. of Bromine Atoms
5th Digit	No. of Iodine Atoms

Example: Halon 1301 = 1 Carbon   Halon 1211 = 1 Carbon
3 Fluorine   2 Fluorine
0 Chlorine   1 Chlorine
1 Bromine   1 Bromine
0 Iodine   0 Iodine

DRY- AND WET-CHEMICAL SYSTEMS

A dry-chemical extinguishing agent is a mixture of fine powders which have been treated to be water repellent. They can be stored under pressure and discharged through piping or hoses. Dry chemical is primarily intended to combat fires involving liquids, gases, grease and electrical equipment. Some formulations can also be effective against fires in ordinary combustibles such as paper and wood.

The dry-chemical formulations are effective fire extinguishing agents for a number of reasons. One important reason is that they are capable of breaking the chemical chain reaction in flame propagation. They also reduce oxygen in the flame zone and reduce flame feedback radiation. In the case of liquid fires, they reduce the evaporation rate of the liquid.

DRY CHEMICALS

The types of dry chemicals commonly used in systems that are effective against liquid, gas, grease and electrical fires have one of the following chemical agents as their base: monammonium phosphate, sodium bicarbonate, potassium chloride and potassium carbonate, which is a product of the fusion of potassium bicarbonate and urea. They all discharge as a cloud and leave a residue which under most conditions does not adhere to a hot surface. The residue can usually be vacuumed or brushed away after the fire is completely extinguished. These types of dry chemicals are nonfreezing, which permits them to be stored under a wide range of temperature conditions.

System Description

Dry-chemical agents are used in total flooding systems (Figure 2-14), in local application systems (Figures 2-15 and 2-16), and in combinations thereof. These systems can be designed for either manual or automatic actuation. The extinguishing agent may be delivered to the fire through fixed-pipe or hand hoseline systems.

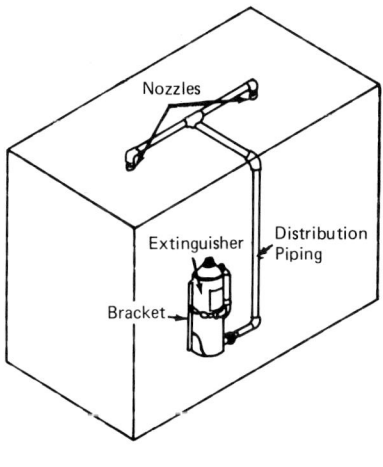

Figure 2-14 A total flooding dry-chemical system uses fixed piping to deliver the agent. *(Courtesy of the Ansul Co.)*

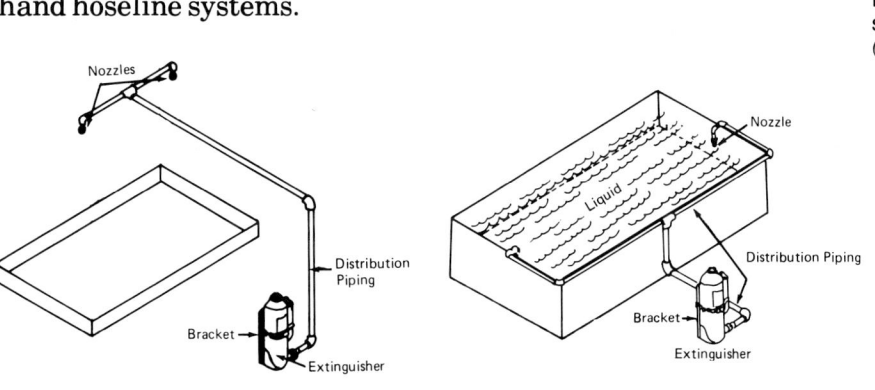

Figures 2-15 and 2-16 Local application systems are designed to blanket an area after manual or automatic initiation. *(Courtesy of the Ansul Co.)*

Common components to all dry-chemical systems are the containers for the agent. The agent and the expellent may be in the same or separate containers. The expellent gas can be either nitrogen or carbon dioxide, depending on the applications. The actuating system and the delivery system through which the dry chemical is discharged are similar regardless of agent used (Figure 2-17).

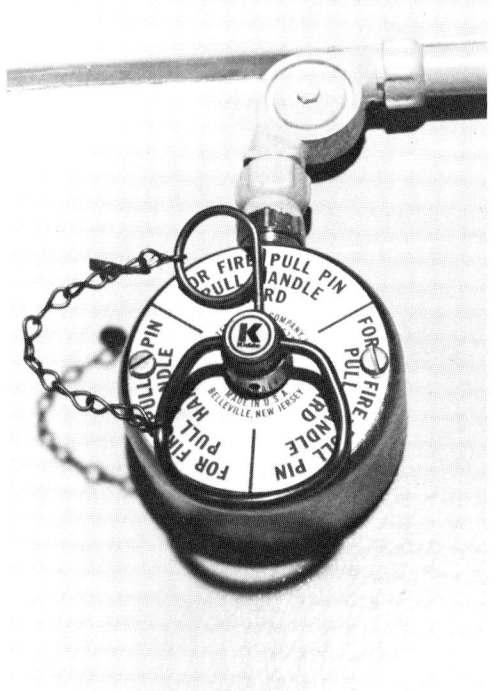

Figure 2-17 A typical actuator, delivery system, and storage container for a total flooding dry-chemical system for use in small installers.

Flow Characteristics

The flow of a dry-chemical mixture with gas does not follow general hydraulic principles because of its two-phase flow, which means a solid material is being transported in a stream of gas. Flow characteristics are dependent upon the composition, physical characteristics, and chemical nature of the type of agent being used, the type and charge of the expellent gas and the design of equipment being used. It is extremely important to follow individual manufacturer's installation recommendations.

System Design

Properly designed dry-chemical fire extinguishing systems assure that adequate quantity, flow rate, nozzle pressure and discharge pattern will be applied to the fire. When these are predetermined by a fire test for a particular type and size of hazard, they are called pre-engineered systems. An engineered system requires individual design and calculation. These take into account piping and nozzle requirements, flow rates and agent quality in order to assure proper protection. Expert assistance may be required to check and approve the adequacy of these systems because of the sophisticated nature.

System Application

Dry chemical is noted for its extremely fast knockdown and extinguishment capability on flammable liquid fires. Therefore, many system applications are found in industrial and petrochemical areas, such as dip tanks, process equipment, loading racks and flammable liquid storage areas.

The nontoxic qualities of dry chemicals have made them extremely popular for use in restaurant hood and duct systems, including protection of equipment such as ranges, fryers and broilers (Figure 2-18). Dry chemical is also nonconductive. While it is usually not used to protect delicate electronic equipment, it is widely used in fires involving motors, generators, transformers and the like.

A recently developed dry-chemical system uses pressurized pneumatic tubing to detect a fire and deliver extinguishing agent. The system is maintained at a predetermined pressure and supervised by a low-air-pressure-warning device. When a fire occurs, the tubing systems and the loss of air pressure causes the dry-chemical storage tank to discharge its contents. This system has broad applications for total flooding of an area such as the engine compartment of bulldozers, crawler tractors and ore haulers (Figure 2-19).

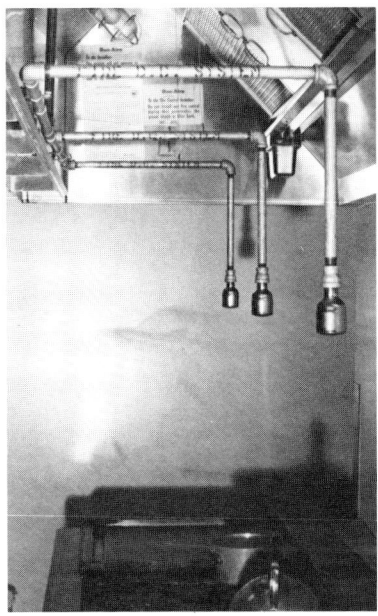

Figure 2-18 Dry-chemical systems are popular for the fire protection of restaurants.

Figure 2-19 A new application of dry-chemical units is on heavy machinery and equipment using pneumatic tubing to deliver the agent.

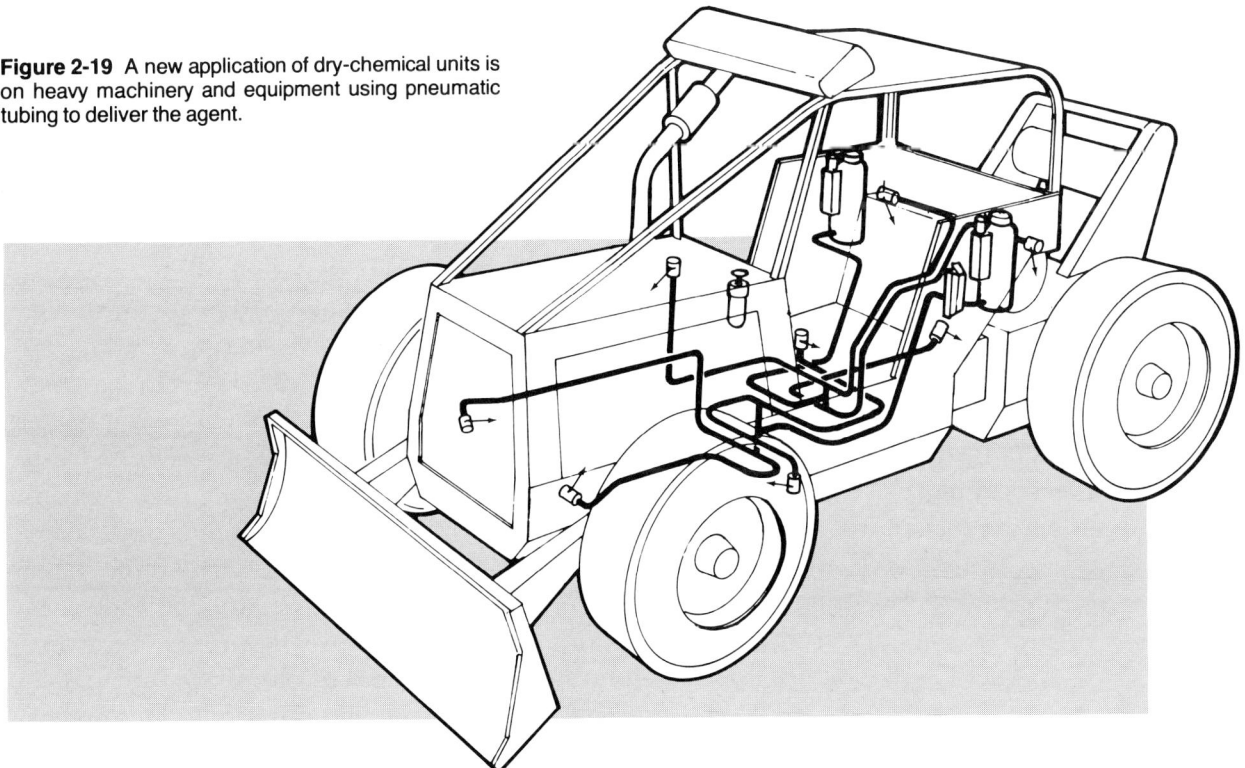

Extinguishing Limitations

Dry chemical is available in several chemical compounds. One compound may be rated for Class B and C fires, while another compound will extinguish Class A, B and C fires, but these cannot be interchanged and maintain a listed system. It is important to take the compound capabilities into consideration when evaluating the hazard to be protected.

Residue is a concern with dry-chemical agents

Residue is a primary consideration when using dry-chemical extinguishing agents on delicate electrical equipment. The powder residue often causes a problem in the electrical equipment after the fire is extinguished.

Another limitation in extinguishing a fire with dry chemicals is reignition after the system has discharged. The burning material may still reignite due to hot spots. Although dry chemicals are good extinguishing agents, their ability to "cool" the material is not considered significant. A water-spray system may be used in conjunction with the dry-chemical system to cool hot spots. Damage may occur when water and a multi-purpose compound combine to form a sticky mass that requires considerable cleanup.

Fixed System

The fixed dry-chemical system can have two methods of application. First, a total flooding application expels the agent into a closed atmosphere. The amount of agent is precalculated to handle the fire load of the hazard to be protected. A dry-chemical system of this type usually actuates the automatic closing devices on vents and openings to the hazardous area. Also, an alarm signal is sounded locally and may be transmitted to the fire department. The second type is called a local application. This application discharges the extinguishing agent directly into the fire. This system is primarily used to protect open tanks or vaults.

Fire Department Operations With Dry Chemicals

Fire department operations, where dry-chemical fixed-extinguishing systems are located, should be planned in advance. The basic principle would be to support whatever extinguishing systems are present. However, there are certain hazards that firefighters should be made aware of. Dry-chemical powders are generally thought to be nontoxic, but when discharged in a closed area certain problems occur. Persons who breathe high concentrations of dry-chemical powder suspended in air may experience coughing and irritation of the respiratory tract. Applications greater than two pounds per second have been found to reduce visibility to potentially dangerous levels. Because of these two factors, positive-pressure self-contained breathing apparatus should be required and a search should be initiated for persons unable to escape from a flooded area.

Breathing and visibility will be impaired by high application rates

One other point of concern, witnessed under certain conditions, what is called a flareup phenomenon. When a large pool of burning fuel is interfaced with dry chemicals there may be a brief release of thermal energy. Firefighters should be made aware of this possible occurence and trained to approach the fire slowly, with full protective clothing.

Hand Hoseline System

The hand hoseline system provides the ability to supply large amounts of extinguishing agents to fires covering large areas such as gasoline loading docks, flammable liquid storage areas and aircraft hangars. The hose stations are connected either directly or by means of intermediate piping to a dry-chemical container as shown in Figure 2-20.

Figure 2-20 Some local application systems use a larger quantity of dry-chemical and a hose station to provide protection. *(Courtcoy of Walter Kidde & Co., Inc.)*

DUAL OR COMBINED AGENTS

The concept of utilizing a combined agent in the extinguishment of large flammable liquid fires was initiated after it was determined that some extinguishing agents were compatible. When it was determined that potassium bicarbonate dry chemical and aqueous film-forming foam were compatible, they were used jointly to extinguish flammable liquid fires, such as aircraft, bulk storage and refineries.

The quick knockdown action of the dry chemical extinguishes the fire, while the foam prevents re-ignition, due to its vapor-securing action. The combined agent apparatus is usually affixed to a vehicle. The agent nozzles can be mounted on the roof of the vehicle where they are moved with a turret-type action by the operator. However, some vehicles are equipped with portable nozzles which are designed for one-man operation (Figure 2-21).

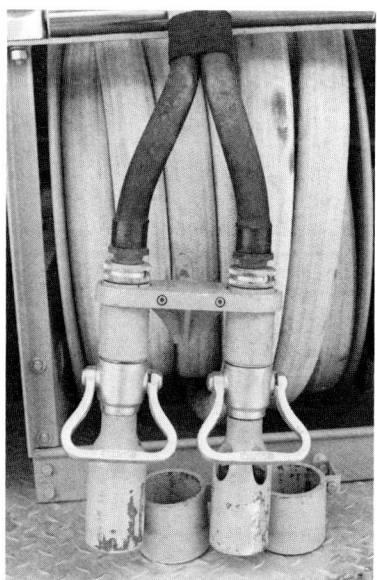

Figure 2-21 Dual-agent units of dry-chemical and aqueous film forming foam have been developed for protection of flammable and combustible liquid storage and use areas.

78 PRIVATE FIRE PROTECTION

Figure 2-22 Wet-chemical systems are used in cooking areas to protect equipment, hoods and ducts. *(Courtesy of "Automatic" Sprinkler Corp.)*

WET CHEMICALS

The extinguishing agent known as wet chemical is an aqueous solution of potassium carbonate. The agent is delivered to the hazard area in the form of a spray. It is an excellent extinguishing agent for fires involving a flammable liquid, gas, grease and ordinary combustibles such as paper and wood. It is not recommended for electrical fires because the spray may act as a conductor.

Wet-chemical systems are most effective on fires caused by cooking hazards. The nature of the chemical is such that it reacts with animal or vegetable oils and forms a soap. The result of using a wet-chemical agent is that grease or oil fires are attached by the four ways in which a fire can be extinguished; namely, fuel removal cooling, smothering and flame inhibition.

Wet-chemical systems are effective on animal and vegetable cooking oils

Removing the Fuel From the Fire

A cooking fire may occur in the deep-fat frying area, the surface of the filters over the cooking area, the walls of the plenum chamber, in the ducts or elsewhere. Fixed nozzles release the wet-chemical agent, which combines with the oil or grease and forms a soap foam permeated with super-heated steam and carbon dioxide. The low-density foam creates a cover over the fat or oil, removing the fuel from the fire by placing a mechanical barrier between them (Figure 2-22). This is called saponification, which is the chemical conversion in which fatty substances combine with an alkali to form a soaplike froth over the surface.

Cooling the Fire Area

When a wet-chemical extinguishing agent comes in contact with hot surfaces in a fire area, the aqueous carrier converts to steam. This steam conversion produces a cooling effect. The cooling is proportional to the level of thermal involvement. The hotter the surface, the greater the quantity of heat that is absorbed and dissipated into the air. This cooling action aids in extinguishment.

Smothering the Fire Area

A wet-chemical extinguishing agent reacting with burning animal and vegetable oils or grease also removes the free oxygen. This is accomplished by the steam generated, and the carbon dioxide created during saponification that dissipates the oxygen. In addition, the saponification creates a barrier as described under fuel removal, which excludes the oxygen from the fuel.

Inhibiting the Flames

Likewise, wet chemicals inhibit the flames by interrupting the chemical chain reaction of fire. The phenomenon of inhibiting flame involves a secondary activity in the flame area. During the burning process, vapors from the heated fuel contain atoms or molecules which remain unchanged during the initial burning process. These liberated particles may have an electrical charge which either attracts or repels other particles. New compounds are created by radicals forming with other atoms and other radicals, or compounds that are present in the reaction zone. This reaction continues throughout the flame area. The extinguishing agent tends to combine with the molecules throughout the flame area to form other molecules which will not unite with oxygen, and thereby breaks the chain reaction.

Automatic Wet-Chemical Systems

As mentioned, wet-chemical systems are used almost exclusively for cooking hazards involving oil and grease used at high temperatures. An automatic system is highly recommended because these are high hazard areas. The system can be pre-engineered based on fire tests for specific types and sizes of hazards, or engineered to a specific hazard situation.

Automatic detectors can be used with bypass controls that are manually operated. Similarly, the actuation system can be so designed that a manual override can be used when the situation requires. The wet-chemical system uses fusible-link and cable automatic releases. Frequently, these systems include an automatic shutoff for the gas or electric supply.

Designing a wet-chemical system for cooking area hazards must be based on the gathering of pertinent information. The number, size, location and type of nozzles are dependent on the size and shape of the hood, duct, plenum and cooking surfaces. The size of the gas line is also important as are the characteristics of the electrical cooking appliances.

INSPECTING AND TESTING CHEMICAL SYSTEMS

In general, all dry- and wet-chemical systems should be tested at least annually. Special conditions which could cause damage to a system warrant more frequent inspections of the system. The expellent gas should be checked semi-annually. In systems with separate expellent gas containers this is accomplished by checking the pressure, if nitrogen is used, or the weight, if CO_2 is used. A list of manufacturer's listed minimums should be obtained. In the stored-pressure-type check the external pressure gauge.

Inspect chemical systems semi-annually

Semiannual inspections should be made of the dry-chemical agent. Either a visual check of the agent level or determination by weight should be made, with the exception of the stored-pressure-type systems. Annual checks should be made to determine if the agent is "caking." Hoselines, nozzles and piping should be checked periodically for damage or obstructions. When attempting to recharge a system, it is imperative that the manufacturer's instructions are followed explicitly.

FOAM SYSTEMS

Foam was first employed as a fire suppressant in the latter part of the nineteenth century. The first foams were chemical-foam agents which were the result of mixing an alkaline salt and an acid salt in water. This reaction resulted in a stable foam forming from the carbon dioxide given off from the chemical reaction. Chemical foams were initially employed against coal oil and other similar hydrocarbon fuels. Today, modern fire fighting foams have been developed for use against the same hazards, while chemical foams have become obsolete.

Foams are well suited for use against flammable liquids because of their light density, high water content, blanketing tendencies and resistance to rapid breakdown. These properties allow foam to be floated over burning flammable liquids, smothering the fire out, and cooling hot objects in and near the flammable liquids. Foams are not suitable for water-reactive or three-dimensional fires, such as leaking flanges.

Do not use foam on water-reactive metals

There are three basic types of foam agents. These are chemical foams, protein foams and synthetic based foams. Of these three the protein and synthetic are generally called mechanical foams to distinguish them from the chemical-type foams. All three foams can be utilized with fresh water or salt water.

CHEMICAL FOAMS

Chemical foams are made when an acid salt and an alkaline salt come into contact in solution. The acid salt is usually termed the "A" ingredient and the alkaline salt called the "B" ingredient. The mixing of the "A" and "B" ingredients to form a foam solution is achieved in three fashions. The first, is by introduction of dry "A" and "B" powders into the water stream through two separate chemical-foam hoppers. The second method is to mix the "A" and "B" powders in the correct proportions and to introduce them into the water stream through a single chemical-foam hopper. A single-hopper chemical porportioner is shown in Figure 2-23. The third method is to premix the "A" and "B" powders with water into an "A" solution and a "B" solution. The two solutions are then pumped together for foam generation.

Systems using chemical foam have inherent maintenance problems due to the corrosive nature of the "A" and "B" powders. Powder mixing often results in clogging problems, and the foams produced have serious drawbacks because of their tendency to "bake" hard and crack. This results in fissures where the flammable liquids can vaporize and burn. For these reasons, and because other suitable foams are readily available, chemical foams are considered obsolete, and included in this section primarily to address remaining installations.

Figure 2-23 A single-foam hopper eductor is used to correctly proportion a mixture of "A" and "B" chemical foam powders in the foam stream.

PROTEIN FOAMS
Regular Protein Foams

Regular protein foams are chemically broken-down natural protein solids. The end product of this chemical digestion is protein polymers, long strands, which have excellent elasticity, water retention capabilities and high strength. These characteristics make protein foam excellent for use as a fire fighting agent. Basic protein foams are nontoxic and work within a 100°F temperature range from 20°F to 120°F. These foams are not well-suited for use on polar solvents, for extremely cold temperatures, for subsurface injection, or for use with dry-chemical powders. However, certain organic solvents can be added to the foam to improve the suitability for use in these situations.

Protein foams are not suited for polar solvents

Regular protein foams, as well as most other foams, are marketed to be mixed into 3 percent and 6 percent foam solutions. This means that in a 3 percent foam solution for every one hundred parts of foam solution (foam concentrate and water mixed) there are 3 parts of concentrate to 97 parts water. For 6 percent solution there are 6 parts concentrate and 94 parts water. The main reason for this difference in concentration is due to technological advancements in foam concentrate production. Initially, the best concentrate available was 6 percent. After a time, procedures were improved to allow a suitable 3 percent foam concentrate to be marketed. The 3 percent gave the advantage of costing less to ship per area of coverage, and requiring only one-half the storage space of 6 percent foam concentrates. Six percent foam concentrates are still marketed to meet the needs of departments that cannot afford to convert their systems to a 3 percent foam concentrate.

Low-temperature foams are protein based foams with added nonflammable antifreeze solutions. These foams can be used at air temperatures down to −20°F. As with normal protein foam, they are marketed in 3 percent and 6 percent solutions.

Fluoroprotein Foams

Fluoroprotein foams are basically protein foams fortified with fluoronated solvents. The addition of these solvents give fluoroprotein foams several advantages over regular protein foams. One of the primary advantages is that these foams tend to separate from the flammable liquids they are mixed into. This quality makes fluoroprotein foams excellent for subsurface injection, or surface applications in which the foam becomes agitated with the flammable liquid. Other advantages are that fluoroprotein foams tend to be more stable than regular protein foams and more compatible for use with dry-chemical agents.

Fluoroprotein foams have several advantages

Fluoroproteins have the same temperature characteristics as regular protein foams. It is also marketed in 3 percent and 6 percent concentrates.

Alcohol Foams

Alcohol foams have been developed for use on polar solvents, such as alcohol, lacquer thinner, acetone and ketones. Polar solvents are miscible in water. This characteristic causes regular foams to break down rapidly. The alcohol foams were developed because regular foams are very miscible in polar solvents, and tended to melt into the burning liquid without extinguishing the fire. Additionally, regular hydrocarbon liquids mixed with even small amounts of polar solvents tended to destroy the effectiveness of regular foam products.

Alcohol foams are not miscible in polar solvents

There is one type of protein-based alcohol foam. This foam is derived from regular protein foam mixed with heavy-metal salts suspended in organic solvents. Protein-based alcohol foams must, to be effective, be applied gently to the burning surface, and must be applied immediately after eduction into the water. Protein-based alcohol foams when pumped through hose or piping lose effectiveness. This loss is in proportion to the distance it is pumped. For this reason, protein-based alcohol foams are usually mixed into the water stream at or near the application nozzle. Application temperature is from 35°F to 120°F.

SYNTHETIC FOAMS
Alcohol Foams

There are two types of synthetic alcohol foams. Both of these foams have the advantage over protein-based alcohol foams in that they do not have to be applied gently to the surface, and can be pumped long distances without loss of effectiveness. Both are marketed in 3 percent and 6 percent solutions and have the same temperature range as protein alcohol foam.

The first type of synthetic alcohol foam is catalytic alcohol foam. It is the result of the mixing of a two-part solution, a polymer and catalyst. The resulting foam is a very stable alcohol-resistant foam. This foam can be pumped long distances without loss of effectiveness.

Synthetic alcohol foam can be pumped long distances

The second synthetic alcohol foam is sometimes referred to as multi-use foam. This is because it can be used effectively on either hydrocarbon or polar liquids, and works effectively when applied through most foam-mixing devices. These foams are synthetically produced concentrates in a single-component foam. The proportioning is very similar to that of regular protein foams, and they may be pumped long distances without loss of effectiveness.

Synthetic Detergent Foams

Several synthetically produced detergent foams are available. These foams have a high foam yield, but are generally less stable than other foams. They have a low resistance to heat or physical destruction, and require high application rates to be effective.

Detergent foams cause a very low surface tension of water, and may be thought of as wetting agents. This makes their use on Class "A" combustibles effective and will sometimes form a frothy emulsion on top of flammable liquids. However, this characteristic causes detergent foams to break down other foams.

Aqueous Film Forming Foams

Aqueous film forming foams (AFFF) are dual-action synthetic foams. The first action of AFFF is similar to the detergent foam's air-entrapping action. This action forms a blanket of strong foam which spreads over the burning liquid surface. This blanket smothers the fire and retards vaporization of the flammable liquid to below flammable limits. The second action is the forming of a film of aqueous solution void of bubbles across the surface of the liquid. This aqueous film smothers the fire and prevents vaporization just as the foam film. The aqueous film is self-healing, and will re-cover open areas caused by agitation of the flammable liquid surface. The result of the foam action and the film forming action make AFFF one of the most dependable and versatile foams available. Use of AFFF foams has the added advantages of not requiring special application devices and being compatible with dry chemical.

AFFF forms a foam blanket and an aqueous film

AFFF foams are generally strengthened by the addition of fluoronated solvents like those used to stabilize protein foams. These foams can be used on flammable liquids, and under some conditions on polar solvents. Often AFFF films are used on flammable liquid spills to prevent ignition of the flammable liquid. AFFF foams are marketed in 3 percent and 6 percent solutions, and have temperature characteristics like alcohol foams.

High Expansion Foams

High expansion foams are foam concentrates mixed into a 2 percent solution and then mixed with air to form high-air-content, good-quality foams. High expansion foam is different from other foams in the respect that it is useful to fight fires in indoor structures and inaccessible places through a total flooding application. It has also been shown to be an effective agent when used with water sprinklers. Outdoor use of high expansion foam can be effective, dependent upon the wind conditions.

High expansion foam can be used for total flooding inside buildings

When using high expansion foams it is necessary to provide ventilation to allow the foam to flow properly. If air becomes trapped ahead of the foam it will not flow adequately.

High expansion foam acts through oxygen displacement in a confined area. It also provides a good insulating blanket which serves as exposure protection. It is nontoxic, but entry into the foam is hazardous because of lack of visibility. Also, foams generated using air contaminated with products of combustion will have the toxic characteristics of the contaminant, and entry will therefore require breathing apparatus. Additionally, the products of combustion tend to cause the foam to break down rapidly.

FOAM PROPORTIONERS

For effective foam generation the foam concentrate must be mixed with the proper ratio of water and then aerated (except, of course, chemical foams and catalytic foams which do not require aeration). It is important that the ratio of concentrate to water is maintained. Too much concentrate wastes the product, and foams lean on concentrate are ineffective.

There are six basic types of recognized foam proportion: foam nozzle, in-line, primary-secondary, pump bypass, metered proportioned and water-motor proportioned.

Foam Nozzle Eductor

The foam nozzle eductor is not common in fixed systems. This is because it requires an eductor at each nozzle. This type of eductor utilizes a venturi action to draft concentrate. It can draft concentrate up to six feet (1.8 m). Immediately after the venturi proportionment the foam-making solution is aerated by the nozzle. It is well-suited for use with protein alcohol foams which need to be applied soon after mixing. The foam nozzle eduction is a one location application appliance because of the trouble of moving the nozzle about along with the concentrate container. The use of a foam nozzle eductor is illustrated in Figure 2-24.

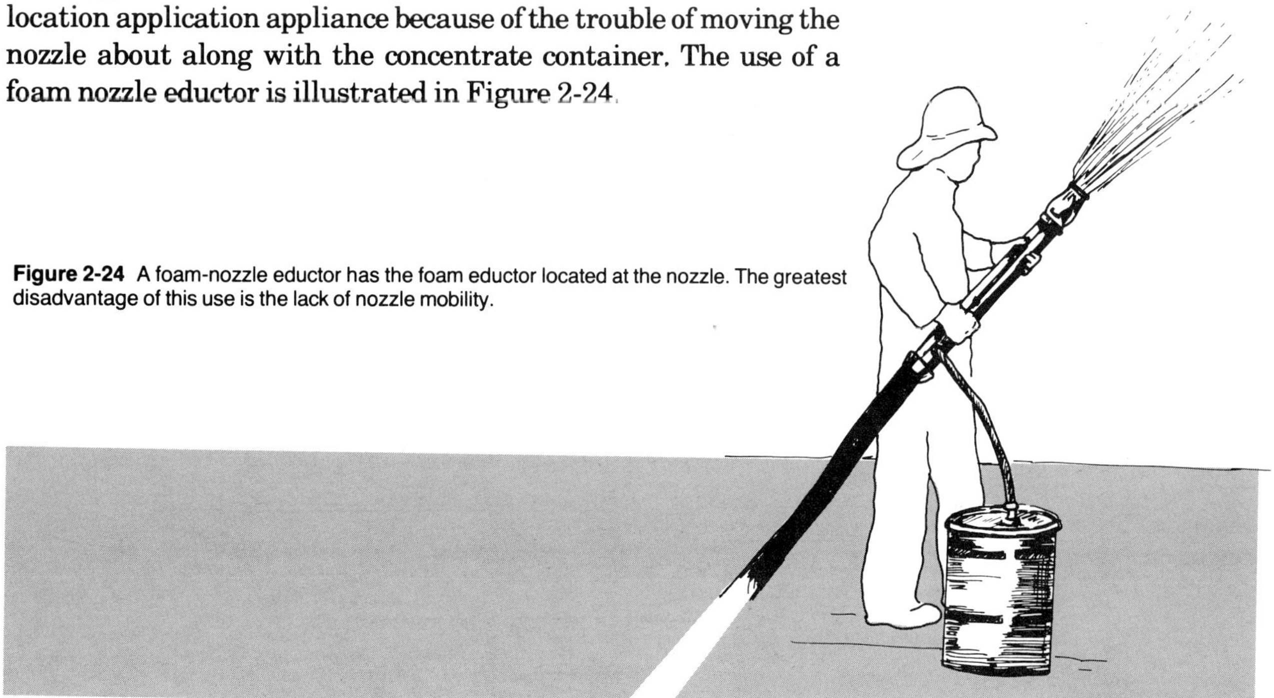

Figure 2-24 A foam-nozzle eductor has the foam eductor located at the nozzle. The greatest disadvantage of this use is the lack of nozzle mobility.

In-Line Eductor

The in-line eductor utilizes the same venturi action as the foam nozzle eductor. However, the foam solution can be pumped to a remote location for aeration. The in-line eductor can be connected directly to a foam nozzle to form a foam nozzle eductor, or can be put on the hoseline or even on the discharge of the pump. Eductor placement should follow the manufacturer's recommendations to allow for the back pressure that will be developed. One eductor can be used to supply more than one line. Caution is necessary when using an in-line eductor in this fashion because this type of device is usually designed to flow a certain amount of water at a given pressure. With multiple nozzles, this design criterion may be exceeded and the foam produced becomes ineffective. Care must be taken to match the nozzle or nozzle combination GPM with the edutor design criterion. Some fixed systems utilize in-line eductors, but most systems depend on more accurate and dependable proportioner systems.

Primary-Secondary Eductor

One common application of the in-line eductor on a fixed system is the primary-secondary eductor arrangement. It has, in fact, two eductors. A primary-secondary eductor is shown in Figure 2-25. The primary eductor is an in-line venturi action eductor. The primary eductor is situated off of a tee from the main waterflow, and water passes through it to the secondary eductor,

Figure 2-25 A primary-secondary eductor is used to produce homogenous foam mixtures when large flows against moderate back pressures are needed.

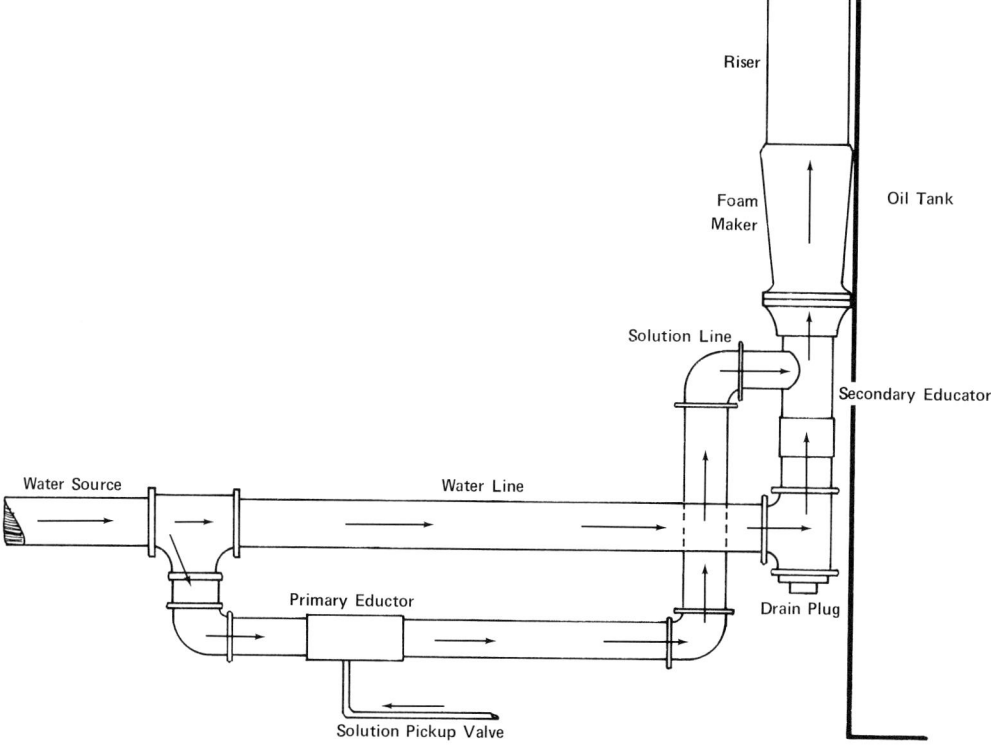

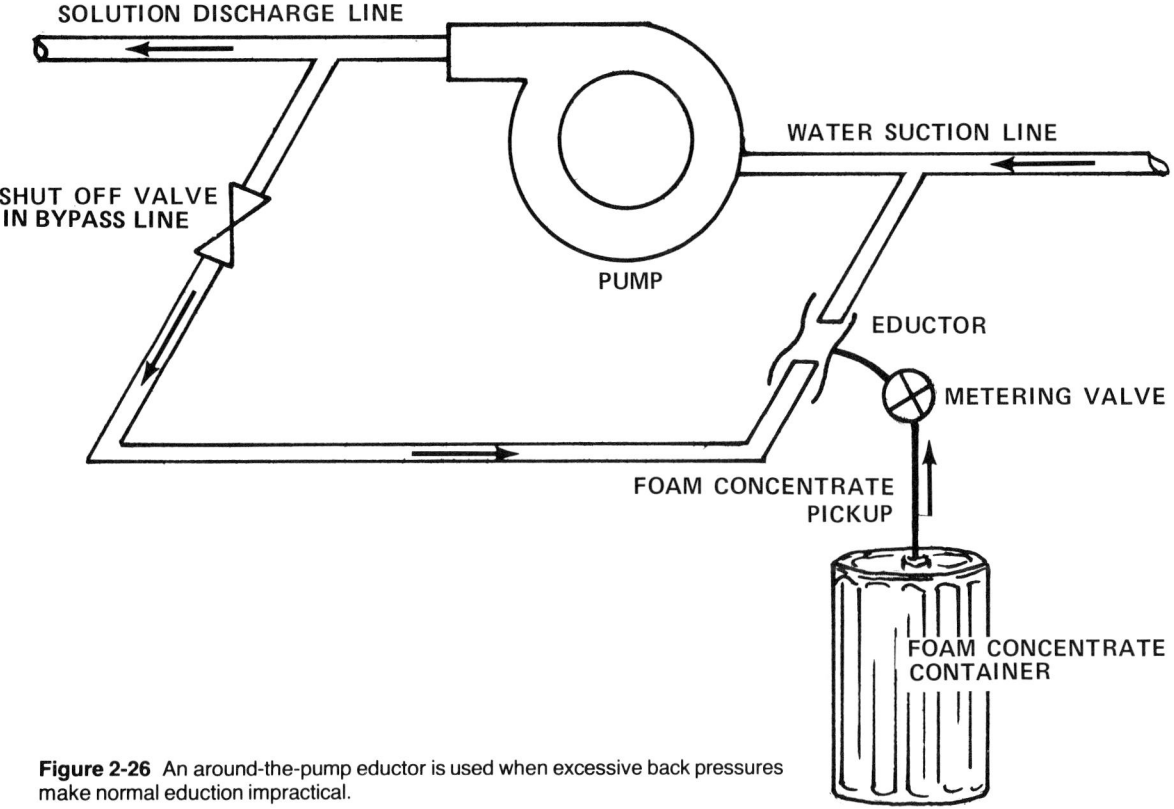

Figure 2-26 An around-the-pump eductor is used when excessive back pressures make normal eduction impractical.

which is located on the main waterflow stream. Eduction at the secondary eductor is a positive-pressure process. The primary-secondary system of eduction is better suited for high-volume foam systems than the in-line eductor. The foam concentrate ratio is better proportioned and mixed.

Around-the-Pump Eduction

The around-the-pump eduction process is typically a fixed-system process. This system runs a pick-up line from the discharge side of the pump back to the suction side of the pump. An in-line eductor is positioned on the pump bypass. The flow of water through the bypass and flow of concentrate through the eductor can be monitored and adjusted in this way. This system is utilized to overcome excessive back pressures in systems requiring the foam solution to be pumped for a long distance, or to an elevation difference causing excessive back pressures. An around-the-pump eduction system is illustrated in Figure 2-26.

Metered Proportioning

Metered proportioning systems deliver foam concentrate into the water system under pressure. These systems utilize a mechanical pump to pump the foam concentrate into the water. The foam concentrate is supplied at a pressure above the normal

88 PRIVATE FIRE PROTECTION

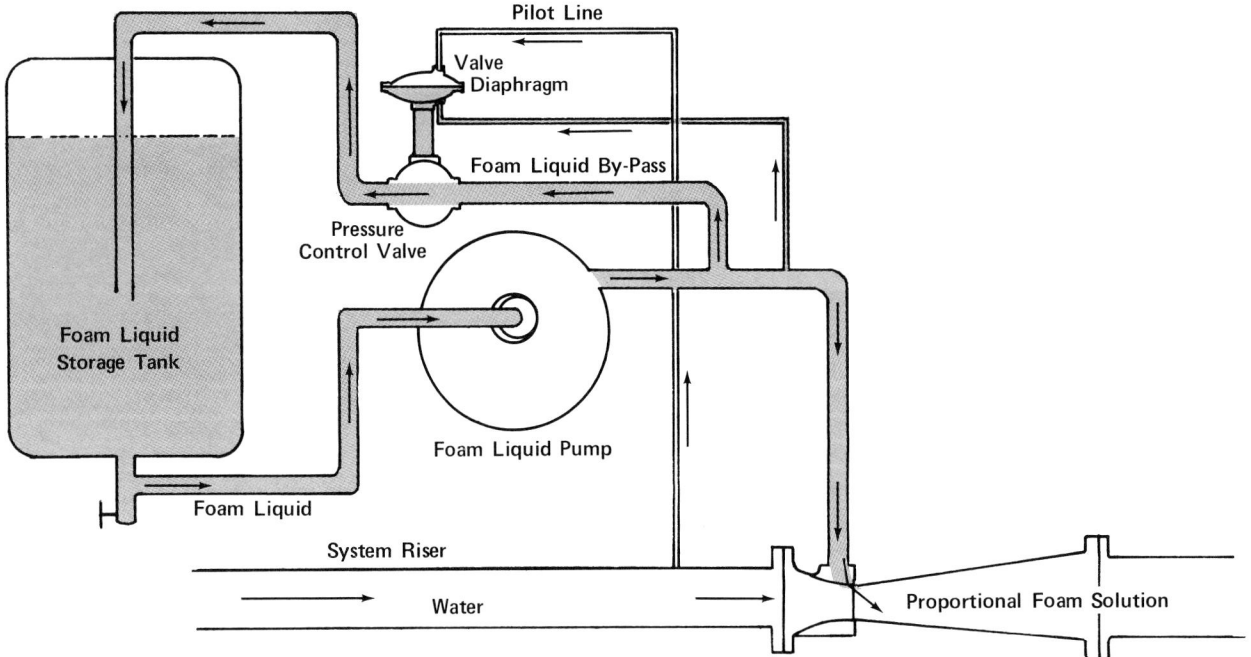

Figure 2-27 A pump-proportioned foam system provides the most reliable form of foam concentrate mixing.

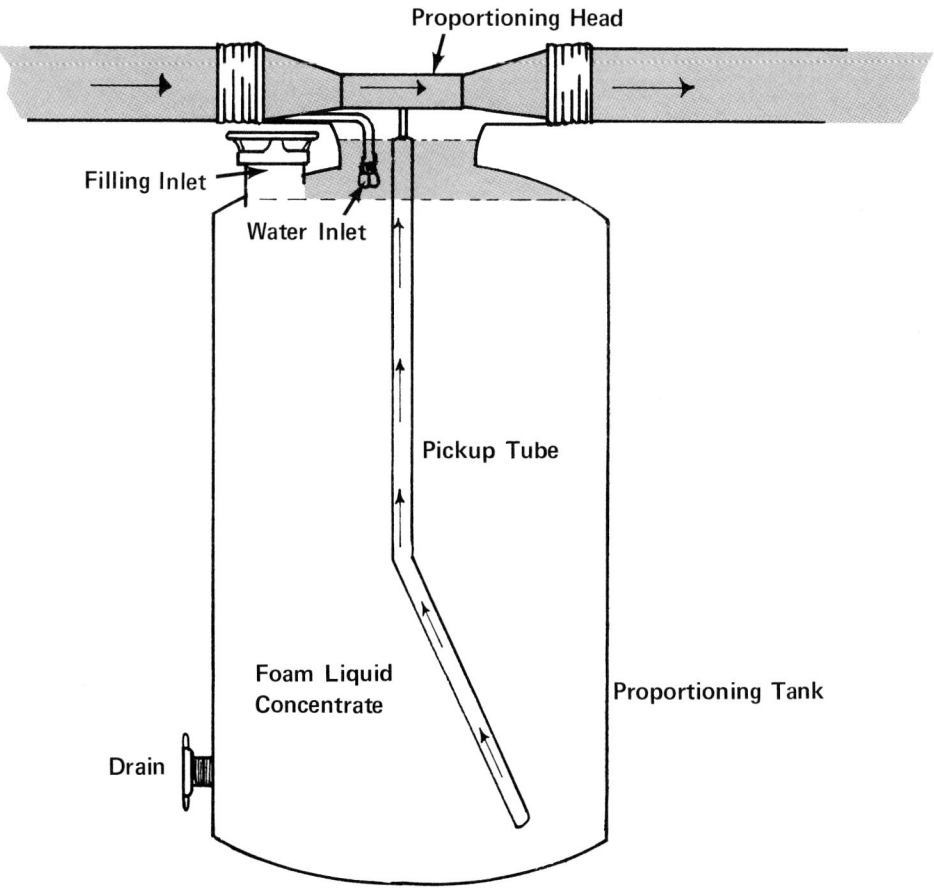

Figure 2-28 A pressure-proportioning mixing system utilizes the velocity of the water and the density differences of the water and foam concentrate to correctly mix the foam solution.

waterflow pressure. The correct proportionment is achieved through a pressure equalization valve. This method is one of the most dependable methods of foam proportionment. A metered proportioning system is shown in Figure 2-27.

Pressure Proportioning

Pressure proportioning utilizes a reservoir of foam concentrate in a tank. The foam concentrate is displaced when water is pumped into the tank. It then flows into the water stream, through a venturi. The two liquids remain separate in the tank due to density differences. This separation is shown in Figure 2-28. The advantage of this system is that it increases the ability of the venturi to educt effectively at high pressures and flow rates. The major disadvantage is that this type cannot be recharged while the system is flowing. The tank must be drained and refilled. For this reason, two or more tanks are usually piped in parallel so one can feed foam while the other is being refilled. It should be noted that some systems use an elastic diaphragm inside the tank, and do not mix the liquids. An illustration and photo of a diaphragm proportioned tank is shown in Figure 2-29.

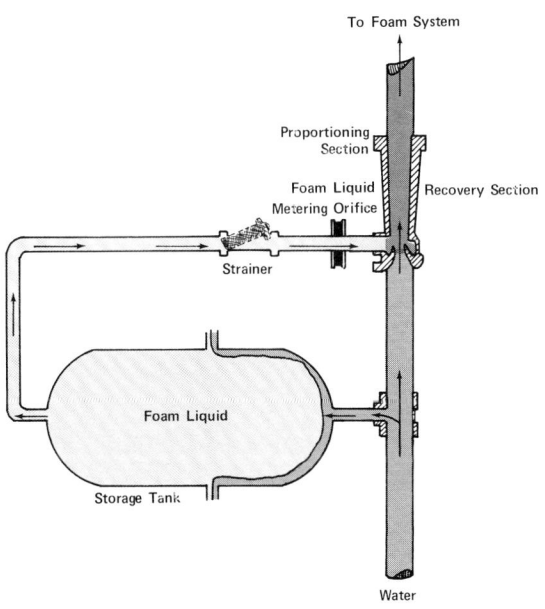

Figure 2-29 A diaphragm-proportioned system utilizes a flexible diaphragm in the tank to keep the foam concentrate and proportioning water separate. The system's major disadvantage is the refill time required. Many such systems are utilized on mobile units. *(Courtesy of Seaway Pipeline, Inc.)*

Water-Motor Proportioning Pump

The water-motor proportioning pump is a hydraulically powered foam concentrate pump. It is, in fact, two positive displacement rotary pumps built back-to-back on a common drive shaft. The size of the impellers in the pumps is in proportion to the ratio of foam to water required. The water side of the pump is inserted into the water stream, and the water flowing through causes rotation. This rotation drives the companion foam concen-

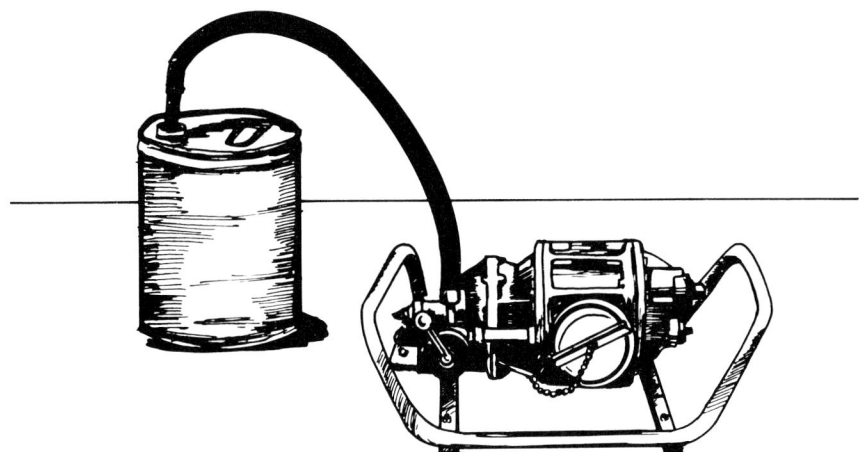

Figure 2-30 A water-motor foam pump correctly proportions foam concentrate into a foam solution at varied waterflows utilizing the waterflow to drive the foam pump.

trate pump via the common drive shaft. The foam concentrate is proportioned correctly with the water flowing at a range of flows from 60 to 180 GPM. There is a larger model which can handle flows up to 1,000 GPM. A water-motor proportioner is shown in Figure 2-30.

FOAM GENERATORS

The process of mechanical foam production can be explained using the foam tetrahedron. Four components are required to produce quality foam applications: water, foam concentrate, air and mechanical aeration. Two combinations of these elements is shown in Figure 2-31. The proportioning devices mix the water and foam concentrate into a foam-making solution which is delivered to the foam generator. The foam generator then combines the foam-making solution and air through mechanical aeration.

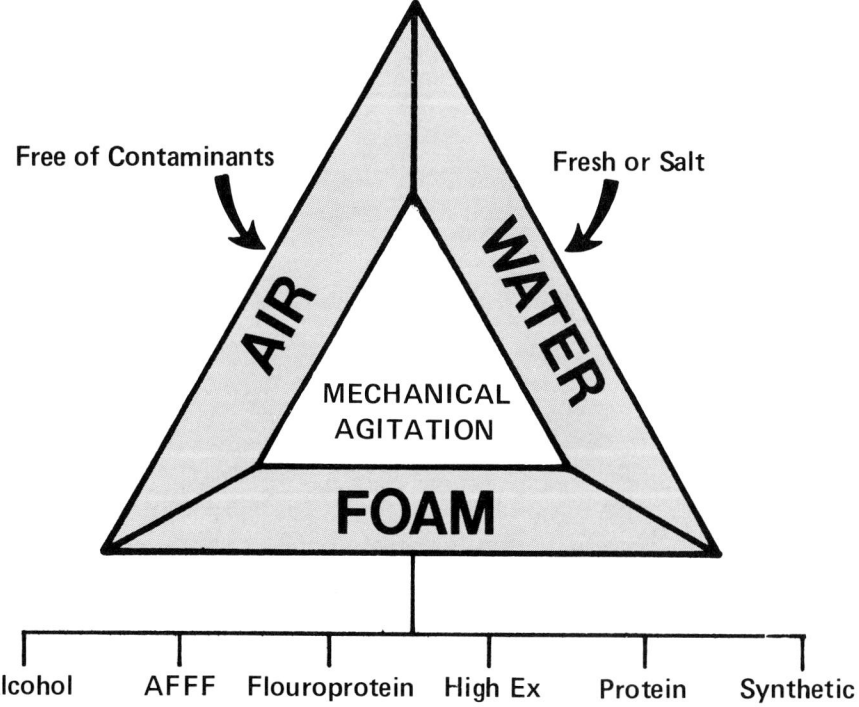

Figure 2-31 The foam tetrahedron represents the four basic requirements of good-quality foams. These are air, water, foam concentrate and mechanical agitation.

Special Extinguishing Systems

Foam generators fall into two major categories: foam nozzles, which includes high expansion generators, and in-line foam generating devices. The difference between the two groups is that in-line foam generating devices produce foam from foam-making solution and the foam is then pumped to a remote location for distribution, and foam nozzles are application devices which produce foam at the end of the line immediately before application. Foam nozzles constitute the most common group of foam generating devices, and the in-line generators are not typically found except in very large systems.

Foam Nozzles

Foam nozzles are foam application devices. These nozzles can be adjusted or moved to change the foam application pattern. One type of foam nozzle has already been discussed, and this was the foam nozzle eductor. This particular nozzle is special as it has the eductor built into the nozzle. Other foam nozzles require a separate eductor. The remainder of the foam nozzles fall into three categories: foam sprinklers, foam-aspirating nozzles and high expansion foam generators.

Foam sprinklers, aspirating nozzles and high expansion generators are types of foam nozzles

Foam Sprinklers

Foam sprinklers are found on foam deluge and foam water systems. Foam sprinklers resemble aspirating nozzles in that they utilize a venturi velocity to mix air into the foam-making solution. However, some systems utilize standard sprinklers to form a low-grade foam through simple turbulence of the water droplets falling through the air. These systems generally use AFFF foams.

Foam sprinklers come in upright and pendant designs. The deflector on them must be adapted to meet the specific installation requirements. Typical sprinklers are shown in Figure 2-32.

Foam-Aspirating Nozzles

Foam nozzles come in two major categories: the fog-foam nozzle and the variable-pattern fog nozzle. These nozzles are employed on fixed systems for manual application of foam in

Figure 2.32 Specialized sprinkler heads produce the highest quality foam on fixed foam extinguishing systems.

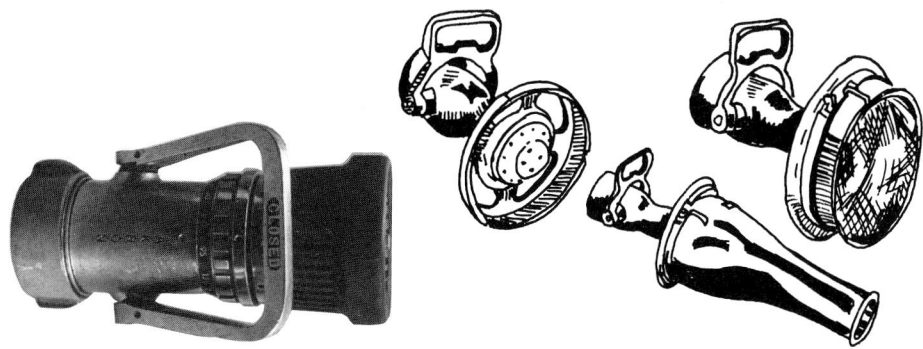

Figure 2-33 A fog nozzle can produce suitable foams through the agitation of the finely divided water droplets with air. These nozzles are particularly well-suited for use with AFFF.

conjunction with the system application. They give the system the added versatility of using portable nozzles. The fog-foam-type nozzle is marketed with two stream-shaping foam adapters. Types of foam nozzles are shown in Figure 2-33. The basic nozzle breaks the foam-making solution into small streams, plus inducts air through a venturi action. The attachments serve to give the foam application qualities different than the standard nozzle. The cone-shaped attachment allows the nozzle extra reach, and the screen produces a more homogeneous high-air-content-foam for gentle applications.

The water fog nozzle, as used with foam solution, produces a low-quality short-standing foam. This nozzle breaks the foam solution into tiny droplets and utilizes the agitation of water droplets moving through air to achieve the foaming action. Its best application is when coupled with AFFF, which, due to their filming characteristics, do not require a high-quality foam to be effective. Water fog nozzles have found a growing acceptance because of the capability to be used as both a normal fire fighting nozzle and a foam nozzle (Figure 2-34).

High Back Pressure Aspirators

The high back pressure aspirator, or forcing foam aspirator, is an in-line aspirator utilized in situations requiring foam to be delivered under pressure. These devices are best suited for sub-surface injection, but are used in conjunction with many fixed-piping systems. High back pressure aspirators operate through venturi action. This action typically produces a low-air-content but homogeneous and stable foam. A high back pressure aspirator is pictured in Figure 2-35.

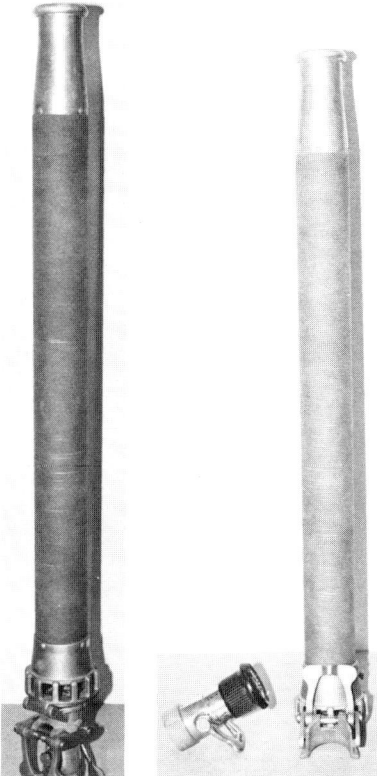

Figure 2-34 The addition of a foam-generating device to a water fog nozzle increases the expansion ratio and quality of the foam blanket.

Figure 2-35 A high back pressure foam aspirator is used in fixed systems to mix foam solution and air under moderate pressures.

Foam Pumps

Foam pumps are positive-displacement pumps with the suction side provided with air ports. The pump is connected to a foam solution supply which mixes the air as it enters the suction side of the pump. Foam pumps are used in situations requiring that foam be delivered under pressure. The foam produced is of a good quality and very stable.

High Expansion Foam Generators

High expansion-foam generators produce a high-air-content, stable foam. The air content ranges from 100 parts air to one part foam solution (100×) to 1,000 parts air to one part foam solution (1,000×). There are two basic types of high expansion foam generators: the mechanical blower and the water aspirating. The water-aspirating is very similar to other foam-producing nozzles except it is much larger and longer. The back of the nozzle is open to allow airflow. The foam solution is pumped through the nozzle in a fine spray which mixes with air to form a moderate expansion foam. The end of the nozzle has a screen, or series of screens, which breaks the foam up and further mixes it with air. These nozzles typically produce a lower-air-volume foam than do mechanical blower generators. A cutaway of a high expansion nozzle is shown in Figure 2-36.

Mechanical blower generators resemble smoke ejectors in appearance (Figure 2-37). They operate along the same principle as the water-aspirating nozzle except the air is forced through the foam spray instead of being pulled through by the water movement. These devices produce a higher-air-content foam, and are typically associated with total flooding applications.

Figure 2-37 Portable high expansion foam generators look similar to mechanical smoke ejectors. *(Courtesy of Walter Kidde & Co., Inc.)*

Figure 2-36 High expansion foam nozzles utilize the energy of the moving water to agitate the foam solution and air. The foam solution is then forced through a fine screen at the end of the nozzle to form a high expansion foam.

FIXED FOAM SYSTEMS

Fixed foam systems are permanently mounted extinguishing systems provided to protect a specific hazard. They may be complete, including automatic detection, activation and foam delivery, or simply be the piping for fire departments to attach their foam equipment. No matter what type of system, the first-responding companies must have firsthand knowledge of the system operation.

Semi-Fixed Systems

Semi-fixed systems are of two types. The Type A, in which the foam discharge piping is in place but is not attached to a permanent source of foam, and the type B. The Type A semi-fixed system requires a separate source for foam solution. This type of system is found in settings which involve several similar hazards. Each is provided with the basic piping distribution system for application. Then a mobile foam solution apparatus is maintained to respond in the event of a fire. These systems can be compared to dry standpipe systems. A Type A system is shown in Figure 2-38 and a portable unit shown in Figure 2-39.

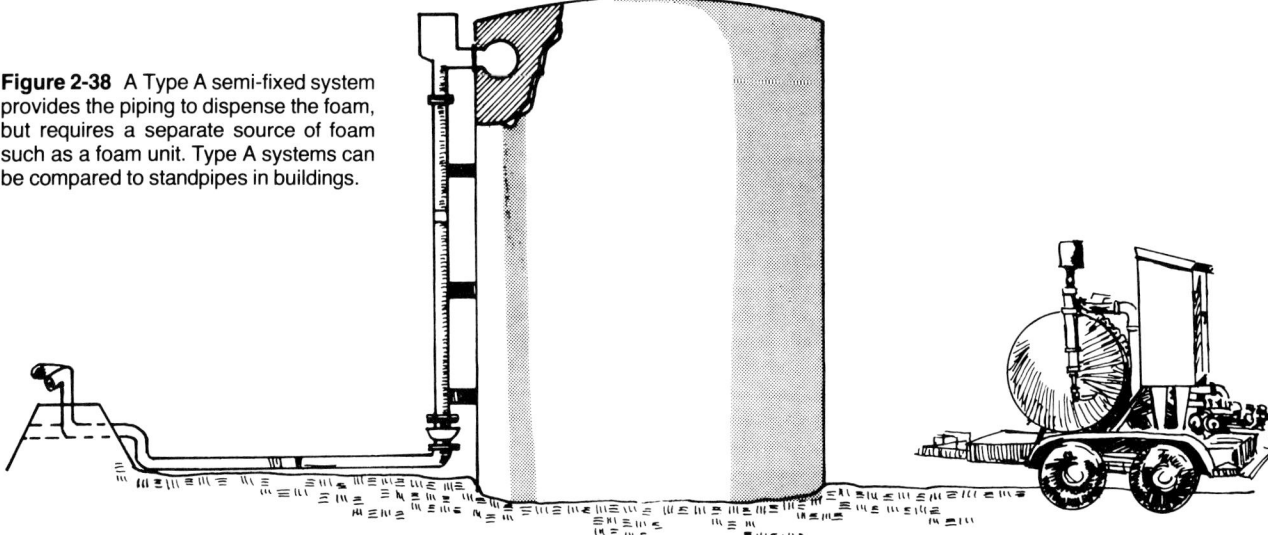

Figure 2-38 A Type A semi-fixed system provides the piping to dispense the foam, but requires a separate source of foam such as a foam unit. Type A systems can be compared to standpipes in buildings.

Figure 2-39 A portable unit to deliver foam solution to a Type A semi-fixed system can be used to cover a number of different installations. *(Courtesy of Seaway Pipeline, Inc.)*

Special Extinguishing Systems

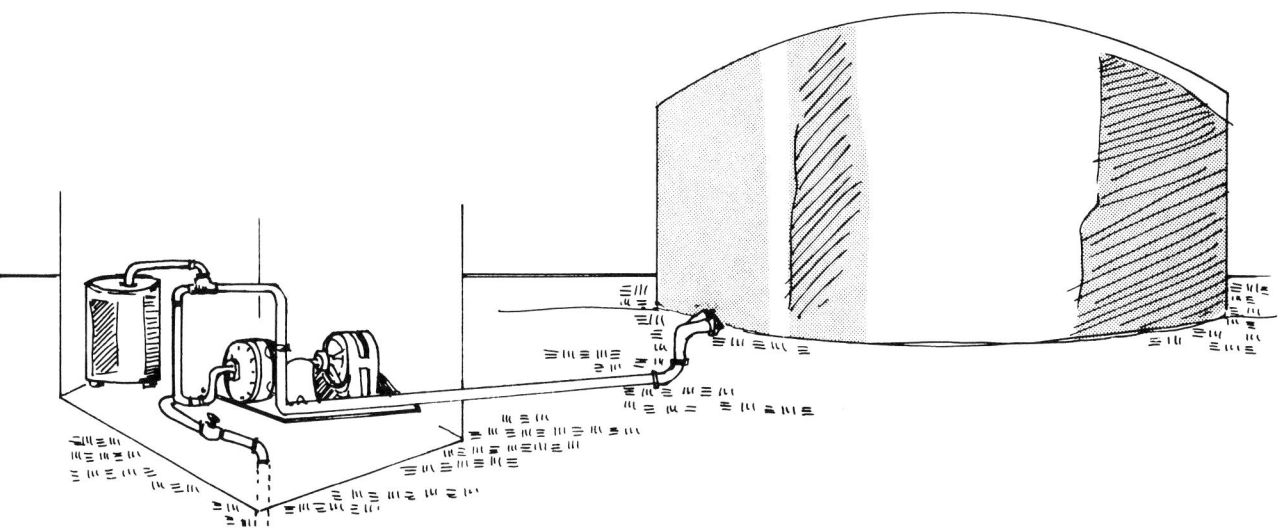

Figure 2-40 Semi-fixed Type B systems have a permanent source of foam concentrate, but require separate foam application devices.

The Type B semi-fixed system provides a foam solution source which is piped throughout the location much like a water distribution system. The foam solution is delivered to foam hydrants for connection to portable foam application devices. A Type B system is illustrated in Figure 2-40.

Foam Monitor Nozzles

Foam monitor nozzles are often used with both Type A and B semi-fixed systems. Foam monitors are large-volume foam nozzles. They are typically aspirating-type nozzles, but some are designed for application of pumped foam. Generally, foam monitors are provided a foam-making solution, but some have eductors as an option. However, because of the large-volume characteristic of foam nozzles, this is the less desirable method.

Monitor nozzles range from 250 GPM to 4,000 GPM. These nozzles can be fixed-position nozzles, manually direct, automatically oscillated or remotely controlled. Foam monitors are most commonly used for protecting tank farms, loading racks and tanker ship piers.

Foam Towers

Foam towers are portable foam-application appliances, although some may be found in fixed applications. Use of a foam tower is illustrated in Figure 2-41. Foam towers are most often utilized in conjunction with Type B semi-fixed foam systems, but may be utilized in conjunction with portable foam-producing units. Some have been truck-mounted similar to an aerial ladder. These devices are very dangerous because they require men to approach the tank.

Figure 2-41 A foam tower is a telescoping device utilized to pump foam over the top edge of a tank from the ground.

Fire department aerial equipment has been used as foam towers by placing a foam nozzle on the tip. However, this requires the use of a high back pressure eductor and a foam pump, or foam concentrate to be educted at the nozzle. Other alternatives may be innovated, but they must be field-tested to determine the results. Also, it is recommended that the manufacturers be contacted concerning the use of their equipment.

Outdoor-Tank Fixed Outlets

Because the approach to outdoor storage tank fires often involves tremendous risk to firefighters, tanks are usually equipped with fixed foam-application nozzles. These nozzles can be part of a semi-fixed system or be part of a total foam protection system. There are two types of tank outlets: Type I and Type II. Type I outlets provide for the gentle application of foam onto the surface of the burning liquid with a minimum of mixing of burning fuel with the surface. Type II outlets provide foam in a manner which generally is not gentle and necessarily involves some mixing of the foam with the burning fuel. From these two descriptions it is obvious that protein foams are best matched with Type I outlets and fluoroprotein foams and aqueous film forming foams with Type II.

Subsurface Injection

Subsurface injection outlets are Type II outlets. Their use is a result of large-diameter storage tanks which make it difficult to apply foam to the center of the burning tanks. The effect of subsurface injection on large-diameter tanks is much bettr than surface applications. Subsurface injection is limited to AFFF and fluoroprotein foams. Also, subsurface injection has not proved effective against polar solvents. A subsurface injection hook up is shown in Figure 2-42.

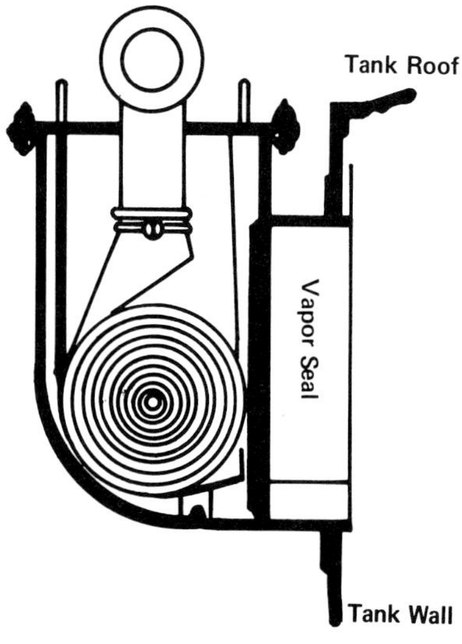

Figure 2-42 Subsurface injection systems can be complete or semi-fixed as pictured here. The system pumps foam to the bottom of the tank from where it floats to the top and spreads evenly. extinguishing the fire. *(Courtesy of Seaway Pipeline, Inc.)*

Moeller Tube Outlets

Moeller tube outlets are Type I outlets. They consist of a porous asbestos tube connected to a foam supply. The tube is kept rolled up in a sealed chamber near the tank (Figure 2-43). The inside of the chamber is a thin diaphragm which can be easily pushed out by the foam pressure. The tube rolls out onto the liquid surface. The foam being pumped through the tube makes it float. The foam flows through the porous tube and onto the fuel surface.

Foam Trough

A foam trough is a fixed metal trough down and around the inside of the tank. A foam outlet that delivers foam to the trough is located at the top of the trough. The foam flows down to the level of the liquid in the tank (Figure 2-44). This type of application is considered a Type I application.

Figure 2-43 A Moeller tube is an asbestos tube which is utilized on tank storage. It rolls out onto the top of the burning liquid when foam is pumped into it. The foam causes the tube to float on top of the oil as it seeps through. This is a Type I, gentle application device.

Special Extinguishing Systems **97**

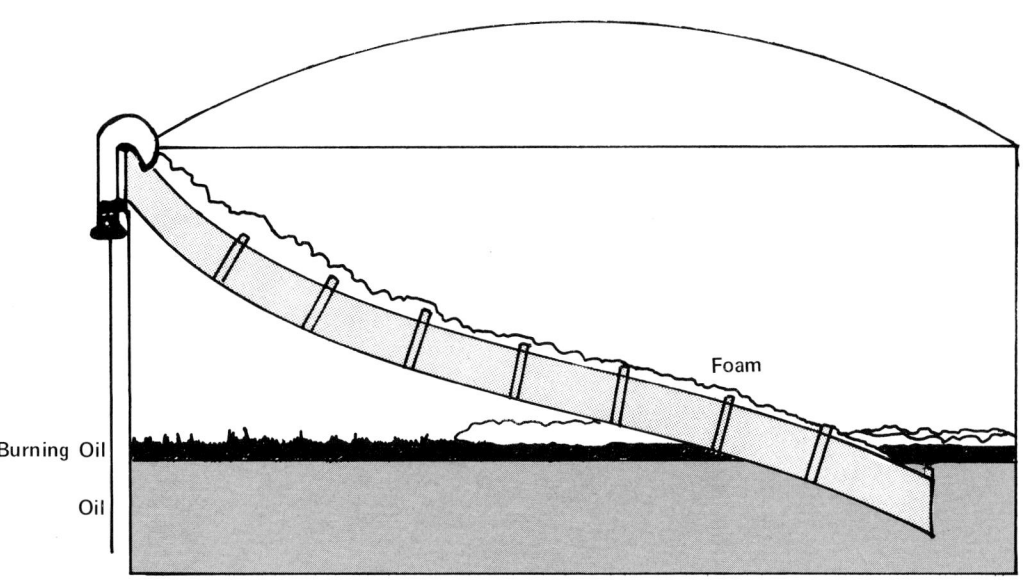

Figure 2-44 A foam trough is attached around the inside of the tank at an angle. The foam travels down the trough to the level of the burning liquid. This is a Type I application.

Side Baffle Outlet

A side baffle outlet works similarly to a foam trough except the foam is allowed to flow vertically down the tank. A side baffle outlet is illustrated in Figure 2-45. The foam flows down a tubular chute with stairstep baffles. Under each baffle is an open port which allows foam to flow out when the chute fills up to that level. This also is a Type I application.

Side Dump Outlet

The side dump outlet is an outlet which flows the foam down the side of the tank. The surface tension between the foam and tank side slow the descent. A picture of a side dump chamber is shown in Figure 2-46. This outlet is a Type II outlet.

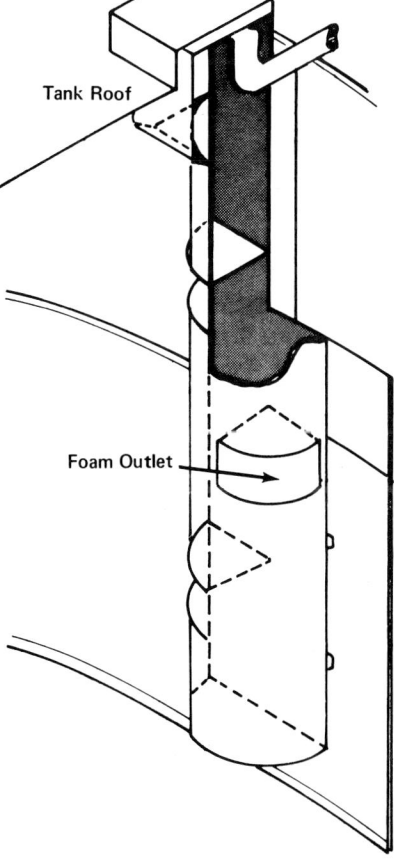

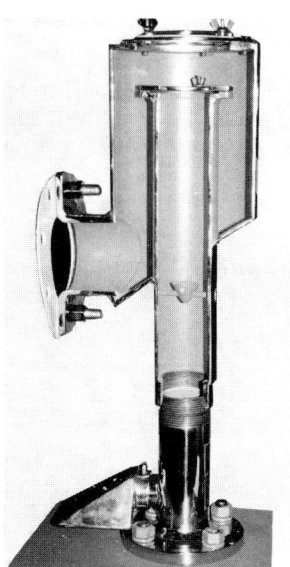

Figure 2-46 A side dump is a chamber attached to the side of a tank the foam flows through the chamber and down the inside of the tank. This is a Type II application device.

Figure 2-45 A side baffle is a tubular chamber attached to the side of the tank. Ports are cut at regular intervals to allow the foam to gently flow out. The foam flows down onto the liquid surface inside the tube, then fills up to and out of the next higher baffle port. This is a Type I application device.

Catenary System

Catenary refers to the delivery device utilized on floating-roof-type tanks. For this reason, the outlet must be capable of moving with the roof. Therefore, a system is employed which combines an outlet on a flexible hose, and fixed piping in a catenary or chain link system.

Foam Walls

Foam walls or foam dams are not application devices but are an integral part of the foam fixed system on floating-roof tanks. On these tanks the fire will be located around the outside edge of the roof; therefore, this is the only place the foam needs to be applied. To achieve this a wall is built around the outside edge of the roof, and the foam outlet fixed to dump foam between the roof and tank wall covering the seal. The foam flows around the seal, extinguishing the fire. A floating-roof-tank seal and foam wall assembly is detailed in Figure 2-47. If the foam system is inoperable, the fire department will have to control the side fire with handlines or dry-chemical while keeping the tank wall cool.

FOAM-WATER SYSTEMS

A foam-water system is basically a deluge sprinkler system with foam introduced into it. It is used in applications where there is a limited foam supply, but an unlimited water supply. Thus, if the foam supply is depleted, the system will continue to function as a conventional deluge sprinkler system. A diagram of a representative foam-water system can be examined in Figure 2-48.

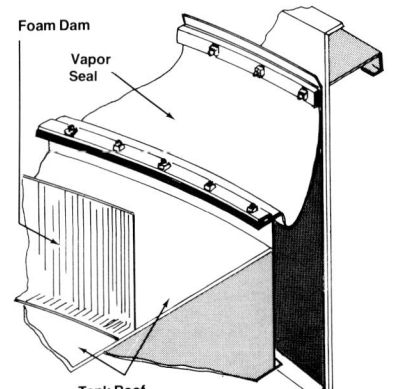

Figure 2-47 The seal between the tank wall and the floating roof can be protected by a foam system which floods the seal area. In order to do this efficiently, foam dams or walls are necessary on the floating roof.

Figure 2-48 A foam-water system combines the versatility of foam with the abundance of water. A system has been developed to use a deluge type system to apply water when the foam concentrate is exhausted.

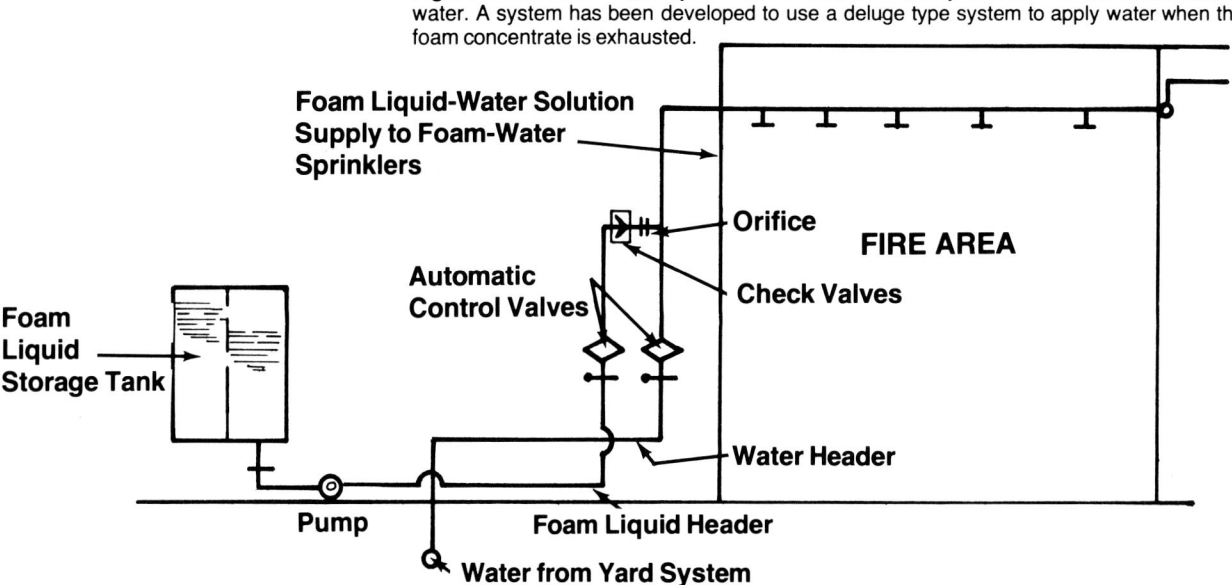

It is an automatic system like a deluge system. The major difference is the use of an air induction type of head, known as a foam head. The system produces a lean foam solution expanded at the head six to eight times to produce a fluid type of foam that will flow around obstructions and even into drains. The most general application is in aircraft hangars.

The entire foam-water system may be divided into two parts: the water system and the foam system. Components of the mechanical foam system used with the sprinkler system consist of a foam liquid storage tank, a metering valve, pump, strainer, piping and actuation unit. Most foam-water systems are deluge types, where the foam is permanently piped into the system, using special open-type, combination foam/water deluge sprinkler heads. The piping arrangement and theory of operation is very similar to the foam-water systems used on foam-equipped apparatus.

The foam sprinkler heads are designed to distribute air (mechanical) foam or water. Foam-water sprinklers are intended to provide an extra margin of protection and are applicable to deluge systems where large quantities of flammable liquids may exist.

In the operation of the system, the heat-responsive devices detect the fire and signal the actuation units for the deluge valves. The sprinkler deluge valve opens, permitting water to enter the system. At the same time, the companion deluge valve opens and the electric motor and foam pump start to introduce foam liquid into the system. The water/foam solution in the piping continues on to the sprinkler heads, where the air is introduced into the solution, producing the proper foam extinguishing agent.

Foam-water systems are activated by heat-responsive detectors

Mechanical foam of the protein, fluoroprotein and aqueous film forming types can be used. Aqueous film forming foam discharged through a standard sprinkler head has equal or better results than through a foam-water sprinkler head. When discharged through a standard head, the aqueous film forming foam solution has a greater velocity which tends to improve the spray and penetration.

After the fire is out, the system is shut down and drained. The foam storage tanks are refilled through the foam filler line. The system is flushed through the open heads or through the flush line. Once the foam system is filled and flushed, the system can be restored to normal. To do this reset the actuation unit and deluge valves.

The procedures for testing and inspecting foam-water extinguishing systems will be the same as the other sprinkler systems previously covered. For additional information see NFPA Standard No. 16, *Foam-Water Sprinkler and Spray Systems.*

HIGH EXPANSION FOAM SYSTEMS

High expansion foam systems are designed for local application or total flooding in commercial and industrial applications. They consist of the following primary components: an automatic detection or manual actuation system (Figure 2-48), the foam generator (Figure 2-50) and the piping from the water supply and foam concentrate storage tank to the generator (Figure 2-51). The foam is generally a detergent base which expands 1,000 to 1 when aerated (Figure 2-52).

These sytems can be activated by any of the common fire detection devices, by a manual pull station, or by both, which is common. The foam generators are powered by electric or gasoline motors, or by water. They should have a fresh-air intake and venting ahead of the foam to allow it to move through the area to be protected. Generators are commonly installed on walls or roofs, but can be ceiling mounted.

Figure 2-49 A high expansion foam system can be supervised through any standard detection system. Note the manual-pull capability. *(Courtesy of Walter Kidde & Co., Inc.)*

Figure 2-50 Fixed systems utilizing high expansion foam generators are arranged to take air from a fresh source. Usually the generators will be placed near or on the roof to allow the foam to flow downward. *(Courtesy of Walter Kidde & Co., Inc.)*

High expansion units can be used to protect Class A or Class B materials since the foam excludes the oxygen from the surface area. Installations include warehouses, storage areas, assembly buildings and even shipboard. They can be installed with automatic sprinkler systems, but are not compatible with dry chemicals or protein foam. Systems are designed to cover the protected area or hazard under two feet of high expansion foam.

Fire departments should be aware of all high expansion systems within their jurisdiction and the major operational features. Although a firefighter can breathe in the material, self-contained breathing apparatus is recommended. The major hazard encountered is the lack of visibility in the foam-filled area. Extreme care must be taken to avoid injuries when working in the foam. Rescue operations cannot only be difficult but hazardous. A path can be cut in the foam with a coarse water spray.

Additional information on these systems is available in NFPA Standard No. 11A, *High Expansion Foam Systems*.

Figure 2-51 Foam concentrate is proportioned into the high expansion foam system through UL listed supply valves.

Figure 2-52 High expansion foam is used for total flooding in indoor structures. In this picture, a rolled paper storage protection system is tested. Note the 24 feet measured on the depth gauge in the center of the photo. The picture was taken after four minutes. *(Courtesy of Walter Kidde & Co., Inc.)*

FOAM APPLICATIONS

Successful fixed-system applications can be described as "The correct foam agent in the correct amount applied by the correct method at the correct time."

The Correct Foam Agent

Determining the best foam agent is usually an easy job when dealing with fixed systems. The reasons for this ease are that the hazard is known before the system is installed and the hazard does not change dramatically. The engineer matches the agent with the hazard in the design stages. Hydrocarbon hazards are matched with protein or fluoroprotein foams, Class A combustibles or flammable liquids are matched with high expansion foams, and polar solvents hazards matched with alcohol foams.

The limitations of fixed foam systems are the same as the limitations of the foams themselves. These limitations are: foams do not work exceptionally well on three dimensional hazards; foams cannot be used on water reactive metals.

Foam does not work well on three dimensional fires

The Correct Amount

Of all the factors involved in a foam system's effectiveness, the foam application rate is perhaps the most important. If the foam is destroyed faster than it is put on. it cannot be effective. The critical application rate is the lowest rate which can achieve extinguishment. The recommended rate is a rate determined by testing and found to be best suited in terms of extinguishment and economy of the system. The engineer of the system will generally be able to supply the specifics of these rates.

The Correct Method

The method of application may or may not be a matter for examination or adjustment. The application may be through fixed nozzles or through adjustable or manual nozzles. If adjustment can be made, some general rules of application are:

- The more gentle the application, the more effective the extinguishment is going to be.
- Before a system should be augmented with other systems or agents, the compatability of the agents must be examined.
- The foam make-up air must be nondestructive to the stability the foam. This usually requires a source of fresh air for the foam generator.

Foam applications should be gentle

The Correct Time

Generally speaking, the correct time is as soon after ignition as possible. Care must be taken when the burning liquid has been allowed a long preburn. If the liquid is hotter than 212°F, the application of the foam may result in a froth which increases the volume and size of the fire without helping to control the fire.

STANDPIPE and FIRE EXTINGUISHER SYSTEMS

- Standpipe Systems
- Fire Extinguishers

The following requirements of the Firefighter Professional Qualifications are met by Section 3:

NFPA STANDARD 1001
Firefighter II

4–16 Portable Extinguishers.

4–16.2 The firefighter shall identify the classification of types of fire as they relate to the use of portable extinguishers.

4–16.3 The firefighter, given a group of differing extinguishers, shall demonstrate the appropriate extinguishers for the various classes of fire.

4–16.4 The firefighter shall identify the portable extinguisher rating system.

5–6 Inspection.

5–6.2 The firefighter shall demonstrate that fire extinguishers in an inspected premises are of required types and ratings, conform to fire prevention code requirements where applicable, and have been inspected and serviced within the required period.

5–6.8 The firefighter shall demonstrate the inspection of standpipe systems for fire protection, including visual inspection of hose (where provided), nozzles, hose outlet threads and fire department connections.

Reprinted by permission from NFPA Standard No. 1001, *Standard for Fire Fighter Professional Qualifications.* Copyright ©1974, National Fire Protection Association, Boston, MA.

STANDPIPE SYSTEMS

Standpipe systems are primarily for the purpose of providing fire stations at each floor level. These stations may be located at various places on each floor and they may also be located on roofs of buildings. Likewise, there are several types of standpipe systems which are classified and discussed in this chapter. A standpipe system must be supplied with adequate water and pressure in order to be effective. Properly installed systems provide a quick and convenient means for operating fire streams on the various floors of buildings. These standpipe stations are essentially indoor fire hose connections. If the system is of the type which permits fire department pumpers to supply the indoor network of pipe, the various stations function similar to discharge ports from a pumper. Proper pressure must be maintained at the pumper and this pressure must be calculated for the highest floor of operation. Other types of standpipes may be equipped with small fire hose and nozzles for use by occupants of buildings as a first defense fire fighting medium. The small hose and nozzle, for this purpose, are kept attached to the standpipe outlet that may be supplied by either a direct connection to the domestic water supply, a fire department pumper or both.

Standpipes provide fire stations on each floor

CLASSES OF STANDPIPE SYSTEMS

The various standpipe systems are grouped into three general classes of service-based upon the intended use.

Class I: For use by fire departments and those trained in handling heavy fire streams (2½-inch hose).

This class of service must be capable of furnishing the effective fire streams required during the more advanced stages of fire on the inside of buildings or for exposure fire protection (Figure 3-1).

Figure 3-1 A Class I standpipe for fire department use in developing a heavy fire stream.

Class II: For use primarily by the building occupants until the arrival of the fire department (small hose).

This class of service must afford a ready means for the control of incipient fires by the occupants of a building during working hours and by watchmen and those present during the nighttime and holidays (Figure 3-2).

Class III: For use by fire department personnel and those trained in handling heavy streams or by the building occupants.

This class of service must be capable of furnishing firefighters the effective fire streams required during the more advanced stages of fire on the inside of buildings as well as providing a ready means for the control of fires by the occupants of the building (Figure 3-3).

In addition to the above, there also exists a combined system, where the water piping serves both 2½-inch outlets for fire department use and outlets for automatic sprinklers.

Figure 3-2 A Class II standpipe for occupant use until the fire department arrives.

Figure 3-3 A Class III standpipe which can be used by the occupants or, with the removal of the reducer, by the fire department.

WATER SUPPLY FOR STANDPIPE SYSTEMS

The amount of water required for standpipe systems depends upon the size and number of fire streams that will be needed and probable length of time they will be used. Both of these factors are largely influenced by the condition of the building or plant. When the character of the water supply must be studied and specific conditions such as acceptability, pumps, tanks and sprinkler systems must be considered, NFPA Standard No. 14, *Standard for the Installation of Standpipe and Hose Systems,* should be consulted.

Minimum Supply For Class I Service

Class I service requires a water supply sufficient to provide 500 gpm for at least 30 minutes. The supply must be sufficient to maintain a residual pressure of 65 psi at the topmost outlet with 500 gpm flowing. Where more than one standpipe is required, the minimum supply shall be 500 gpm for the first standpipe and 250 gpm for each additional standpipe for at least 30 minutes.

Minimum Supply For Class II Service

The minimum water supply for Class II service must be sufficient to provide 100 gpm for at least 30 minutes. The supply must be sufficient to maintain a residual pressure of 65 psi at the topmost outlet with 100 gpm flowing.

Minimum Supply For Class III Service

Class III service requires a water supply sufficient to provide the same flow and the same residual pressure for the same time as stipulated for Class I service.

Class I and III systems supply 500 gpm

TYPES OF STANDPIPE SYSTEMS

There are four different types of standpipe systems currently in use as follows:

- Wet standpipe systems that have the water supply valve open and water maintained throughout the system at all times.

- Dry standpipe systems that admit water to the system through the operation of a manually activated, approved remote control device located at each hose station.

- Dry standpipe systems that admit water to the system automatically through the use of approved devices such as dry-pipe valves. This system is activated by the opening of a discharge outlet valve which depletes the air from the system. This type of system is applicable to unheated structures.

- Dry standpipe systems that have no permanent water supply and must be charged by the fire department.

PRESSURE-REDUCING VALVES

Where pressure at the standpipe outlet exceeds 100 psi an approved device should be installed at the outlet to reduce the pressure with the required flow at the outlet to 100 psi. For Class II and III systems this device would not normally be adjustable to provide pressure higher than 100 psi. Where pressure is greater than 150 psi, an appropriate warning sign should be installed.

STANDPIPES IN HIGH-RISE BUILDINGS

The requirements established in NFPA Standard No. 14 specifies that standpipe systems shall be limited to 275 feet of height. The standard further specifies that buildings in excess of 275 feet of height shall be zoned accordingly. The term "zone" is used to signify those upper and lower levels of a building which are serviced by standpipes. It should be remembered that 275 feet is a maximum height and lower-level zones may be less. Example: A 25-story building may be approximately 325 feet in height. The low-level zone may be established somewhat short of 275 feet, provided the high-level zone is not excessive. Therefore, the low-level zone could be approximately 200 feet, while the high-level zone would be 125 feet. In no case, however, shall the two zones exceed 550 feet in height. A typical two-zone standpipe system and an alternative two-zone system are illustrated by line drawings in Figures 3-4 and 3-5.

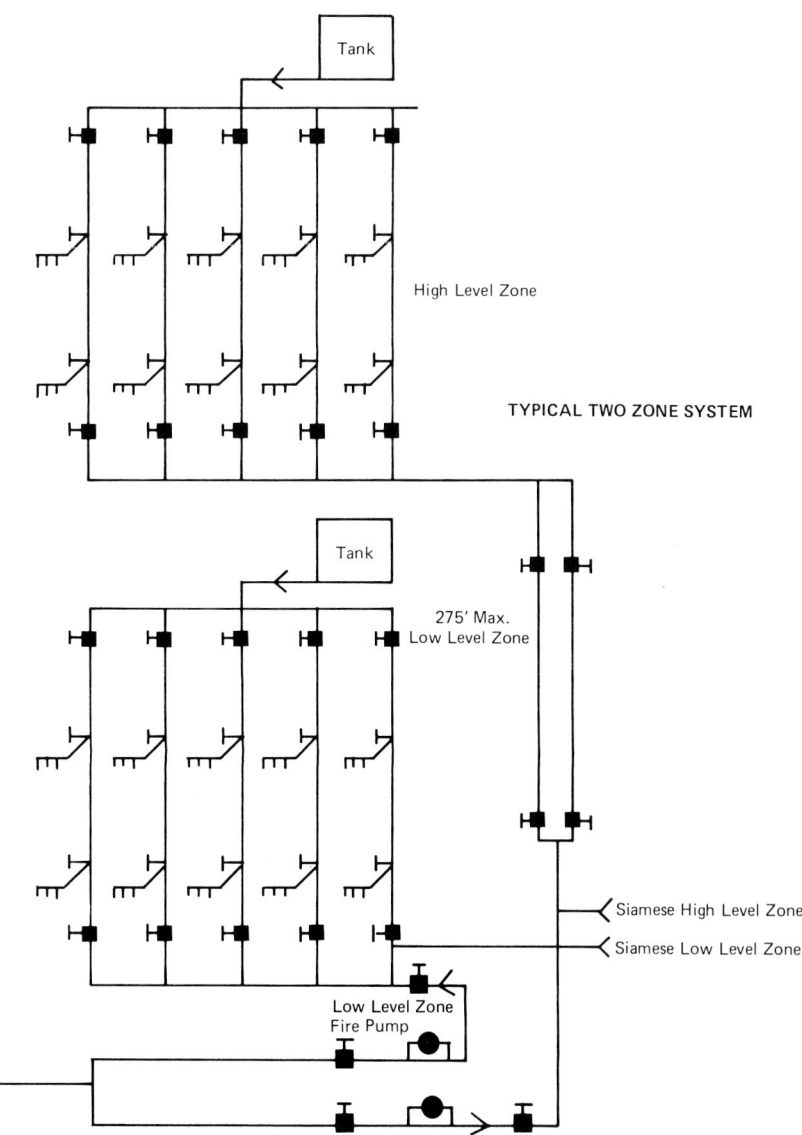

Figure 3-4 A two-zone standpipe system can provide water to floors and a roof up to 550 feet above the ground.

Standpipe and Fire Extinguisher Systems **109**

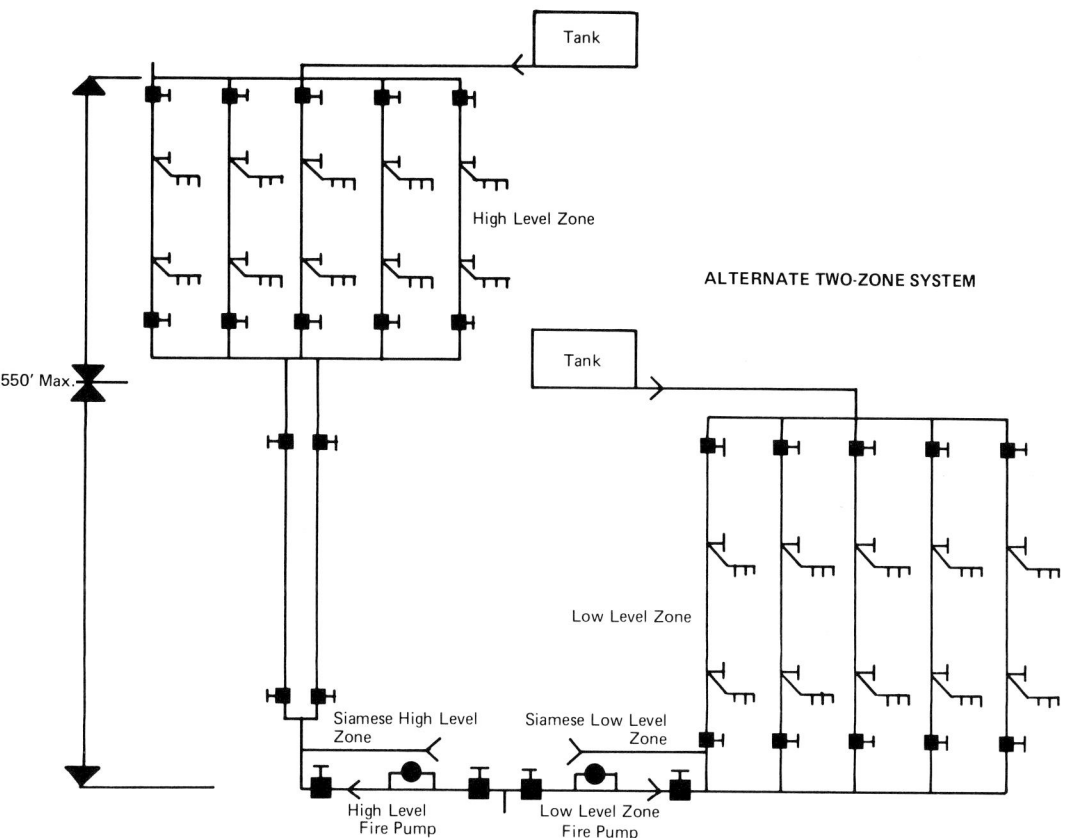

Figure 3-4 An alternative method of supplying a two-zone standpipe system. Fire pumps gravity tanks and fire department connections are utilized.

FIRE DEPARTMENT CONNECTIONS

One or more siamese connections through which a fire department pumper can supply water into a standpipe system is required for each Class I or Class III standpipe system. High-rise buildings having two or more zones require a siamese connection for each zone. Standard requirements specify that there shall be no shutoff valve between the siamese connection and the standpipe riser. An approved straightway check valve, however, must be installed and located as near as practical to the point where the siamese connection joins the system.

The hose connections to the siamese must be of the female type and equipped with standard caps. It is important that the hose-coupling threads conform to those used by the local fire department. The siamese hose connection must be designated by a raised-letter sign on a plate or fitting reading "STANDPIPE." If the siamese connection does not service all of the building an appropriate sign must be attached indicating the portion of the building served. The illustrations in Figure 3-6 show three types of standpipe hose connections.

Figure 3-6 Fire department connections to standpipes may be wall mounted or at a location convenient to the apparatus. By placing a siamese on one side of the connection greater support can be provided.

INSPECTING AND TESTING OF STANDPIPES
Initial Installation

A new standpipe installation should be checked to insure the following:

- All devices are listed by a nationally recognized testing laboratory.
- Hose stations are in no case over six feet from the floor and are within easy reach when standing on the floor.
- Each hose cabinet or closet is provided with a conspicuous sign which reads "FIRE HOSE" and/or "FIRE HOSE FOR USE BY OCCUPANTS OF BUILDING."
- Fire department connections have the proper fire department thread and are posted with a sign reading "STANDPIPE."
- Dry standpipes are posted with "DRY STANDPIPE FOR FIRE DEPARTMENT USE ONLY."
- The system components are tested hydrostatically at not less than 200 psi for two hours.
- Piping, feed mains and connections are flushed to remove all construction debris and trash.

Inspection

An inspector must insure that:

Detailed inspections of standpipe systems must be made regularly

- Wet standpipe systems are inspected at regular intervals with the normal recommendation being every six months. The inspection should include water supply tanks, either pressure- or gravity-type, and fire pumps.
- Hose closets and cabinets are used for fire equipment only.
- Water supply control valves are checked and sealed open.
- Individual discharge outlets are checked for proper operation, gasket condition, presence of corrosion and leaks.
- Discharge outlet threads are inspected for damage.

Note: If the operation of valves in the wet system will be done during the inspection, the main water supply valve should be closed and the system drained before initiating the tests.

- Tanks are checked for proper water level; if pressurized, that at least 75 psi is maintained and that precautions against freezing are taken when necessary.
- Dry standpipe system piping is inspected annually.
- Discharge outlets in a dry system are checked to see they are closed.
- The hose is inspected for general conditions, dryness and proper position on the swingout rack or holder. It should be

removed and reracked periodically so that it will not deteriorate at the bends. The swingout rack should be checked for ease of operation.

- The threads on the combination nozzles and couplings are not damaged and the rubber gaskets are in good condition.
- The combination nozzle is not obstructed and the valve is working freely.
- The fire department connection is inspected for access, protective caps, damage to threads, proper operation of the check valve and that no debris has been placed in the connection which would obstruct the flow of water.
- Drains are checked to be free of dirt or sediment.

Inspect hose cabinets completely

Testing

The following standpipe testing program should be adopted:

- Dry standpipe systems shall be hydrostatically pressure tested at least every five years.
- Any standpipe system out of service for a period of time shall be tested with air at 25 psi to insure tightness.
- Fire pumps supplying standpipe systems should be tested weekly for starting and operating.

FIRE DEPARTMENT OPERATIONS WITH STANDPIPES

Fire departments responding to locations which have standpipes should conduct preplanning surveys in order to develop overall standard operating procedures for the ground level, fire area and all support functions. NFPA Standard No. 13E, *Fire Department Operations in Properties Protected by Sprinkler and Standpipe Systems,* and IFSTA 304, **Fire Problems in High-Rise Buildings** will be of assistance in preparing adequate procedures.

Fire department operations at fires in buildings equipped with standpipes begin at the base of the standpipe where one of the first-arriving engine companies must lay two lines into the siamese connection. This is done to charge a dry system or augment the pressure in a wet system. The pump operator or companies responding to standpipe-equipped buildings must be familiar with the location of the siamese connection and the nearest hydrant in addition to pump operations required to adequately supply fire fighting lines using the standpipe system. Pump discharge pressure will be dependent on the friction loss in the hose lay from the pumper to the siamese, the friction loss in the hose in use on the fire floor, the nozzle pressure for the type of nozzle employed, the quantity of water flowing from the nozzle, the back pressure due to the elevation of the fire above the ground and, in some cases, the friction loss in the standpipe. Generally,

A first-arriving engine company must lay two lines to the siamese connection

Figure 3-7 Standpipe packs are carried on many apparatus to facilitate fire department operations in buildings equipped with standpipes.

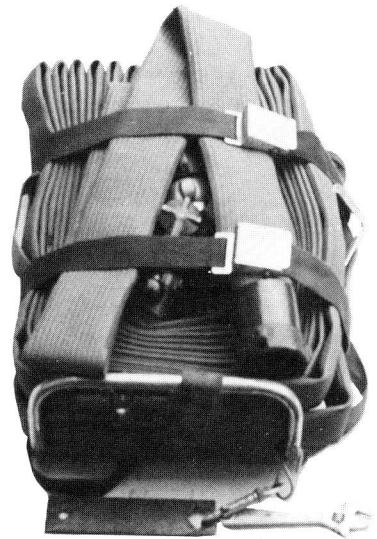

the friction loss in the standpipe is negligible unless the flows are large; however, 25 psi should be added to the pump pressure to allow for losses in the fire department connection, any clapper valve obstruction, pipe bends and hose connections.

A standard operating procedure should be developed so as the company prepares to enter a standpipe-equipped building they take all of the necessary equipment with them. Many departments carry prearranged packs for use with standpipes (Figure 3-7). Equipment which must be considered includes:

150 feet of hose	valve wheel
nozzle	pipe wrench
gated wye 2½ × 1½ × 1½	forcible entry tools
spanner wrench	rope hose tools

In order to reduce the weight involved in this equipment, single-jacketed hose may be used as well as a nozzle and wye constructed of lightweight materials. Some departments are using wheeled handcarts to carry equipment to standpipe connections near the fire area. When wearing breathing apparatus, it may be more convenient to place the standpipe pack in a canvas bag for carrying from the shoulder.

Handlights and a portable radio will also be necessary on the fire floor. Other useful items might be a forcible entry kit and a lifeline. A water thief attached to a short section of 3-inch hose can also be useful for attaching to the standpipe since it allows the use of either 1½- or 2½-inch hose.

There are several advantages to using 2½-inch hose rather than 1½-inch when working from a standpipe. The friction loss is reduced; the fire fighting capability is greater; there is a reserve for the larger fire; with an adjustable gallonage nozzle the flow can be reduced for the smaller fire; the 2½-inch hose will not kink as easily, causing a flow restriction; and usually there are not more than three lengths, which does not make the weight prohibitive.

Before a company leaves the lobby it should review the building floor plan and any special features of the standpipe it will be using. Additional hose and masks should be brought into the building and can be stored in the lobby until needed on the fire floor. Personnel should be assigned to check riser valves, fire pump operations, forcible entry, ventilation and search-and-rescue operations.

Assign personnel to check riser valves

The firefighters should proceed to the floor below the fire. The use of elevators is not recommended because of control and power failure difficulties. They will connect to the standpipe on this floor and stretch the line to the fire floor above. Excess hose should be laid up the stairway toward the floor above the fire. It is much easier to bring the hose down the stairs than up, especially after it has been charged. Care must be exercised so as little obstruction as possible is created for anyone who is exiting the building via the stairwell. Any pressure-reducing device should be removed before the fire department connects to the standpipe riser. One firefighter must remain at the outlet valve to charge the line after it has been moved into operating position. If a larger floor area or large fire is anticipated, it may be best to lay a length of 3-inch from the outlet and place a 2½- × 2½- × 2½-inch wye on it, enabling the use of two hoselines on the fire. The firefighter operating the valve should watch for fire or heat developing behind the attack team and warn them before their position becomes untenable. He should not proceed to assist them on the hoseline until he is certain they will not be entrapped. Additional hoselines in single-riser buildings may be advanced from connections on lower floors if the water supply is adequate.

Lay excess hose up the stairway toward the floor above

The hallway may be extremely hot upon entry and full protective clothing and positive-pressure self-contained breathing apparatus is necessary. Often there will be a lack of ventilation during the initial stages. A straight stream nozzle for fire fighting may create less steam obstructing the fire attack when ventilation is inadequate. Often, fog nozzles are not recommended with standpipes, since they can become easily clogged by sediment in the piping and many standpipes cannot withstand the pressure required to maintain 100 psi at the nozzle.

Whenever a building is so equipped, multiple standpipes should be used for a multiple-line attack. This permits securing of the fire area and protection of internal exposures. Using multiple standpipe lines increases the water supply reliability and enables search and rescue operations in more than one area simultaneously. Multiple lines from different areas provide paths to the various stairwells, giving additional means of egress in times of an emergency.

Often when dry standpipe systems are charged there will be a time delay before the delivery of water due to the amount of air that must be expelled from the system. If they are interconnected, adjacent risers can trap air which may become compressed, putting added stress on the riser. However, cross-connected standpipes in a building with control valves suitably located reduce the effect on the operations if a problem arises with one of the standpipe risers. With multiriser systems it may not be necessary to spend a great deal of time overcoming a standpipe failure. A number of simple operations can be used by the fire department to overcome standpipe difficulties, such as placing a double male and double female on a fire department connection to overcome a frozen swivel, bypassing a connection and entering the riser through the discharge outlet on the first floor by using a double female, or going to an operative discharge outlet below or above where the normal hookup would have been made. To overcome a problem on a single-riser building where the standpipe is totally unserviceable, it may be necessary to hoist a line up the outside of the building. Often this can be accomplished by unrolling donuts from every second or third floor down the side of the building, rather than attempting to hoist the line several floors. Standpipes of adjacent buildings can also be used to protect exposures or aid in fire fighting.

Simple operations will overcome standpipe difficulties

OUTSIDE PRIVATE PROTECTION

Outside private protection is a series of hydrants, located on private property, receiving their water supply through a privately owned distribution system. This distribution system may receive its water supply from the public water supply system or other reliable sources; for example, privately owned tank, or suction from rivers or other natural sources.

The hydrants found on these systems are of two basic types: wall hydrants and yard hydrants. A wall hydrant (Figure 3-8) is basically a pipe protruding through the wall with 2½-inch male connections and a control valve. Yard hydrants may be either the wet-barrel (Figure 3-9) or dry-barrel (Figure 3-10) type, depending on the severity of cold weather.

These systems may also include strategically located hose cabinets (Figure 3-11). These hoselines are intended to be used by private fire brigades and are normally not used by the public fire department. The use of private fire hydrants is dependent upon

Figure 3-8 Outside protection can be provided by a hand-operated wall hydrant.

Figure 3-9 In warmer climates, wet-barrel hydrants can be used as yard hydrants to supply hoselines.

Figure 3-10 Dry-barrel hydrants are commonly found in use as yard hydrants for outside protection.

Figure 3-11 Hose cabinets are often located near private system hydrants to provide equipment for fire brigades.

the design of the yard system and the water supply available. The local public fire department should obtain the system details and limitations from the system owner. This information should be incorporated in the preplan for that location. Additional information can be obtained from NFPA Standard No. 24, *Standard for Outside Protection.*

Outdoor hazards for which hose stream protection is especially needed include: yard storage for combustible materials and equipment, sheds without sprinklers, freight cars, aboveground flammable liquid storage, possible flammable liquid spills, open loading platforms and serious fire exposures from outside the plant property.

Figure 3-12

Green Triangle

Red Square

Blue Circle

Yellow Star

FIRE EXTINGUISHERS

Portable fire extinguishers are not intended to be a substitute for other fire extinguishing systems. Their primary use is as a first-line defense for fires of limited size. They are considered necessary even though the property is equipped with automatic fire protection devices. Additional information on fire extinguishers is available in IFSTA 101, **Forcible Entry, Rope and Portable Extinguisher Practices,** and in NFPA Standard No. 10, *Portable Fire Extinguishers.*

CLASSIFICATION OF FIRES

Since portable fire extinguishers are classified according to their intended use, before the proper type of extinguisher to protect a hazard can be selected, the user must know the type of potential fire that the hazard constitutes. The four classes of fires are:

Class A Fires

Defined as fires involving ordinary combustible materials, such as wood, cloth, or paper, where the "cooling-quenching" effect of quantities of water or solutions containing large percentages of water is most effective in reducing the temperature of the burning material to below its ignition temperature. The symbol for Class A fires is illustrated in Figure 3-12.

Class B Fires

Defined as fires involving flammable liquids, greases and gases where the "smothering-blanketing" effect of oxygen-excluding media is most effective. Other extinguishing methods include removal of fuel, temperature reduction and inhibiting the chemical chain reaction. The symbol for Class B fires is illustrated in Figure 3-12.

Class C Fires

Defined as fires involving energized electrical equipment where the electrical nonconductivity of the extinguishing media is of first importance. The materials involved are either Class A or B and can be handled as such once the equipment is de-energized. The symbol for Class C fires is illustrated in Figure 3-12.

Class D Fires

Defined as fires involving combustible metals, such as magnesium, titanium, zirconium, sodium and potassium. The symbol for Class D fires is illustrated in Figure 3-12.

NEW MARKING SYSTEM

A new extinguisher marking system based on the international "Picture-Symbol" labeling system is now in limited use. The system is designed to make the operation of fire extinguishers more effective and safe to use through the use of the less confusing pictorial labels. The system indicates the type of fire that the extinguisher can be used on, with a light blue symbol background. The new system also emphasizes when NOT to use an extinguisher on certain types of fires. This is indicated by a black background with a red line through the symbol. The regular letters and symbols as shown in Figure 3-12 are also printed on the marker. This is no "Picture-Symbol" for Class D, combustible metal fires. Examples of this labeling system are shown in Figures 3-13 to 3-15.

The quantity of agent must be checked in cartridge-operated dry-chemical extinguishers. This can be done by weighing or by

Figure 3-13 The new type of marking for an extinguisher suitable for Class A, B and C fires.

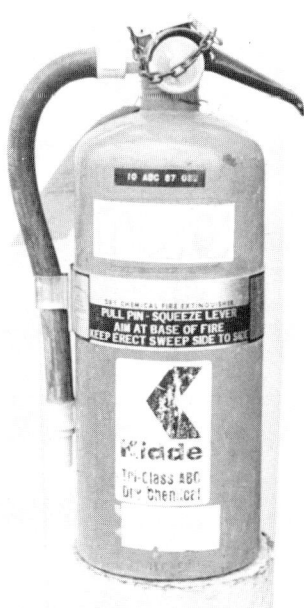

118 PRIVATE FIRE PROTECTION

Figure 3-14 An extinguisher bearing this label is suitable only for Class A fires. This is indicated by the blue background while the backgrounds of the symbols for Class B and C are black and marked with a red "danger" line.

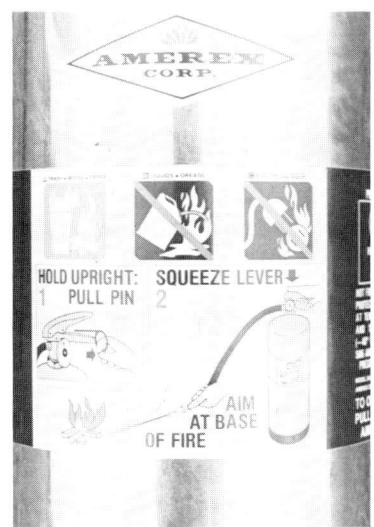

Figure 3-15 An extinguisher with this symbol can be used on either class B or C fires, as indicated by the colors, letters, geometric symbols and pictorials.

RATING OF EXTINGUISHERS

Extinguishers are rated with a letter rating for the class or classes of fire they are designed to control. Class A and B extinguishers also receive a numerical rating that precedes the letter and designates the potential size fire the extinguisher can be expected to suppress. Multiple letters or numeral-letter ratings are used on extinguishers which are effective on more than one class of fire. For example, a 2½-gallon foam extinguisher is rated as 2-A, 4-B. This also indicates that it will extinguish twice as much Class A fire as an extinguisher that is rated as 1-A, or four times as much as an extinguisher that is rated 1-B. It is extremely important the correct extinguishing agent be used on a fire. The use of the wrong agent could be extremely dangerous and could lead to an unsuccessful extinguishment attempt and/or violent reactions.

Class A Extinguishers

Ratings from 1-A through 40-A are designated for Class A fire extinguishers. The number indicates the relative fire extinguishing potential of the extinguisher. One-and-one-fourth gallons of water are required for a 1-A fire. For the ratings 1-A through 6-A an extinguisher must extinguish each of the following types of fires: wood crib, wood panel and excelsior fire (Figure 3-16).

Figure 3-16

Classification & Rating	No. of Members	Normal Size and Length of Members (Inches)	Arrangement of Wood Members
1A	50	2 x 2 x 18¾	10 layers of 5
2A	70	2 x 2 x 24	13 layers of 6
3A	98	2 x 2 x 28¾	14 layers of 7
4A	120	2 x 2 x 31¼	15 layers of 8
6A	153	2 x 2 x 36	17 layers of 9
10A	209	2 x 2 x 44½	19 layers of 11
20A	160	2 x 2 x 60	10 layers of 15 on edge 1 top layer of 10 flat
30A	192	2 x 2 x 72	10 layers of 18 on edge 1 top layer of 12 flat
40A	224	2 x 2 x 84	10 layers of 21 on edge 1 top layer of 14 flat

WOOD CRIB TEST

Classification & Rating	Test Panel Size (Ft.)	Gallons Fuel Oil Applied	Total Pounds Excelsior For Windrows
1A	8 x 8	1	10
2A	10 x 10	2	20
3A	12 x 12	3	30
4A	14 x 14	4	40
6A	17 x 17	6	60

WOOD PANEL TEST

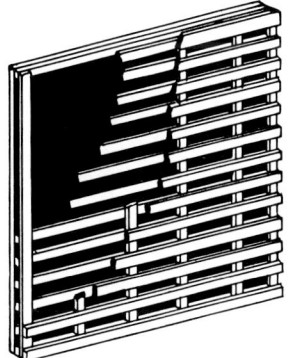

Classification & Rating	Weight of Excelsior (Pounds)	Test Area (Feet & Inches)
1A	6	2' 10" x 5' 8"
2A	12	4' 0" x 8' 0"
3A	18	4' 11" x 9' 9½"
4A	24	6' 0" x 10' 8"
6A	36	6' 11" x 13' 11"

EXCELSIOR TEST

Class B Extinguishers

Extinguishers for use on Class B fires are classified with numerical ratings ranging from 1-B through 640-B. The number is an approximate indication of the area, in square feet, of a fire involving an 8-inch-deep layer of flammable liquid that can be extinguished. For example, a 10-B unit can be expected to extinguish a fire of ten square feet on an 8-inch-deep layer of flammable liquid. This is the rating for an untrained operator. A trained firefighter should be able to extinguish 25 square feet with a 10-B extinguisher (Figure 3-17).

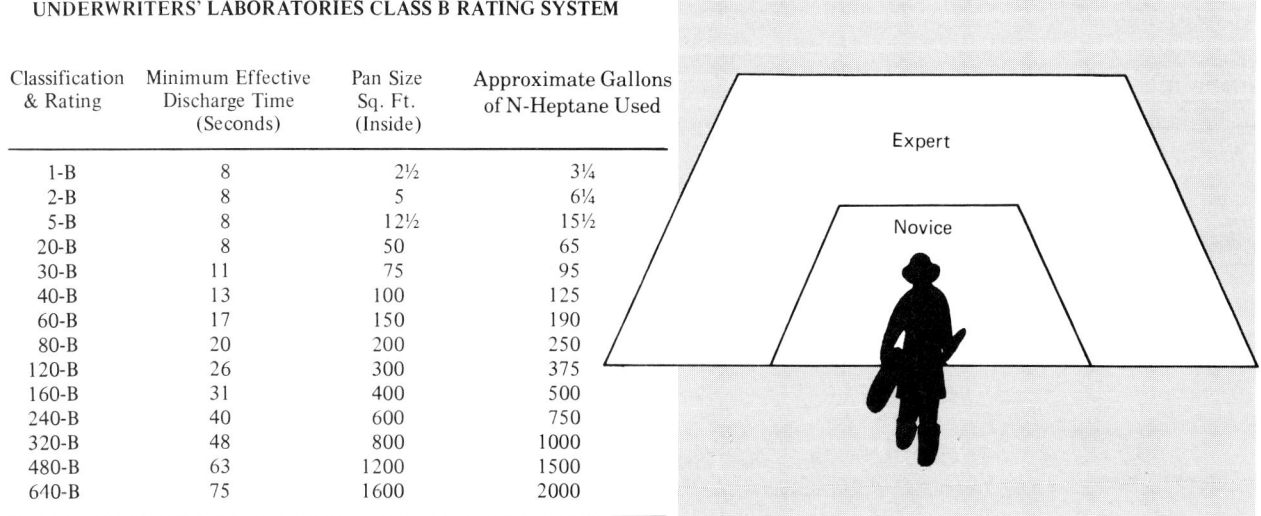

UNDERWRITERS' LABORATORIES CLASS B RATING SYSTEM

Classification & Rating	Minimum Effective Discharge Time (Seconds)	Pan Size Sq. Ft. (Inside)	Approximate Gallons of N-Heptane Used
1-B	8	2½	3¼
2-B	8	5	6¼
5-B	8	12½	15½
20-B	8	50	65
30-B	11	75	95
40-B	13	100	125
60-B	17	150	190
80-B	20	200	250
120-B	26	300	375
160-B	31	400	500
240-B	40	600	750
320-B	48	800	1000
480-B	63	1200	1500
640-B	75	1600	2000

Figure 3-17 A trained firefighter should be able to extinguish 2½ times as much fire with a Class B extinguisher as would an inexperienced individual.

Class C Extinguishers

Extinguishers for Class C fires have no numerical rating and are tested only for electrical nonconductivity. No Class C rating is provided in conjunction with a rating previously established for Class A and/or Class B fires. The size of the extinguisher installed shall be commensurate with the size and extent of the Class A and/or B materials in the electrical equipment or around the electrical hazard.

Class D Extinguishers

Extinguishers for Class D fires have no numerical rating and tests vary depending upon the metal they are intended for. The facepiece of the extinguisher will detail the specific material and instructions for use.

FACTORS INFLUENCING THE EFFECTIVENESS OF FIRE EXTINGUISHERS

In order to determine the number of fire extinguishers that are needed to adequately protect a property, it is first necessary to consider the area and arrangement of the building or occupancy.

It is further necessary to consider the severity of the hazard, the anticipated classes of fires and the distances to be traveled to reach each extinguisher. The anticipated rate of the fire spread, the intensity and rate of heat development, the probable intensity of smoke and the likelihood of a close approach to a fire with fire extinguishers should also be considered. The successful operation of fire extinguishers depend upon the following conditions:

- Extinguisher is properly located.
- Extinguisher is properly placed and in working order.
- Extinguisher is of proper type for the fire which may occur.
- Fire is discovered while still small enough for the extinguisher to be effective.
- Fire is discovered by a person, ready, willing and trained to use the extinguisher.
- Smoke created by burning materials allows approach.
- Fire is accessible for a close approach with portable extinguishers.

For success use a fire extinguisher on a fire small enough to control

DISTRIBUTION OF EXTINGUISHERS

The type, size and number of extinguishers required are based on the occupancy hazards of the area to be protected. The following occupancy classifications are defined in NFPA Standard No. 10.

Light Hazard

A light hazard is a situation where the amount of combustibles or flammable liquids present is such that a fire of small size may be expected. These may include offices, schoolrooms, churches and assembly halls.

Ordinary Hazard

An ordinary hazard is a situation where the amount of combustibles or flammable liquids present is such that fires of moderate size may be expected. These may include mercantile storage and display, auto showrooms, parking garages, light manufacturing, warehouses not classified as extra hazard and school shop areas.

Extra Hazard

An extra hazard is a situation where the amount of combustible material or flammable liquids present is such that fires of severe magnitude may be expected. These may include woodworking, auto repair, aircraft servicing, warehouses with high-piled combustibles, and processes such as flammable liquid handling, painting and dipping.

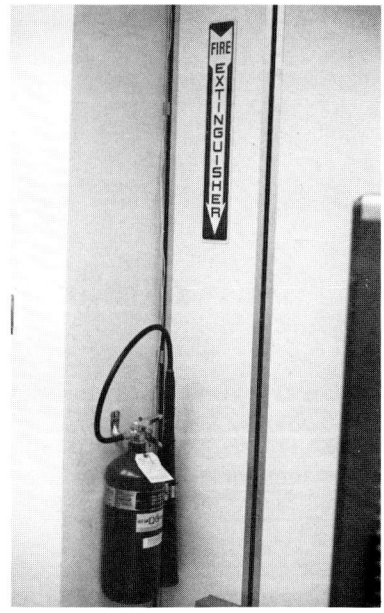

Figure 3-18 Extinguishers should be located so they are accessible and easy to see. To increase their visibility, the location can be marked.

Placement of Extinguishers

The placing of extinguishers is vital to the ease of accessibility. Extinguishers weighing less than 40 pounds should be installed with the top not more than five feet from the floor. Those weighing more than 40 pounds, except wheeled units, are installed with the top not more than 3½ feet from the floor. However, a clearance of at least 4 inches must be maintained between the floor and the bottom of the extinguisher.

Location of Extinguishers

Extinguishers should be placed where they are visible, within reach and available when a fire breaks out. They should be located along normal paths of travel with some extinguishers located near exits from an area (Figure 3-18). Rooms or areas are graded generally as Light, Ordinary or Extra Hazardous.

- Figure 3-19 should be used to determine the placement of Class "A" types.
- Figure 3-20 should be used to determine the placement of Class "B" types.

Basic Minimum Extinguisher Rating For Area	Maximum Travel Distance to Extinguishers	Areas To Be Protected Per Extinguisher		
		Light	Ordinary	Extra
1-A	75 ft.	3000 sq. ft.		
2-A	75 ft.	6000 sq. ft.	3000 sq. ft.	
3-A	75 ft.	9000 sq. ft.	4500 sq. ft.	3000 sq. ft.
4-A	75 ft.	11250 sq. ft.	6000 sq. ft.	4000 sq. ft.
6-A	75 ft.	11250 sq. ft.	9000 sq. ft.	6000 sq. ft.
10-A	75 ft.	11250 sq. ft.*	11250 sq. ft.*	9000 sq. ft.
20-A	75 ft.	11250 sq. ft.*	11250 sq. ft.*	11250 sq. ft.*
40-A	75 ft.	11250 sq. ft.*	11250 sq. ft.*	11250 sq. ft.*

*11250 sq. ft. is considered a practical limit.

- Extinguishers with Class "C" ratings are required where energized electrical equipment may be encountered which would require a nonconducting extinguishing agent. Whenever possible, electrical equipment will be de-energized before attacking Class "C" fires.

- Extinguishers must not be obstructed or obscured from view. Where visual obstruction cannot be completely avoided, means must be provided to conspicuously indicate the location and intended use of extinguishers.

For Extinguishers Labeled Prior to June 1, 1969

Type of Hazard	Basic Minimum Extinguisher Rating	Maximum Travel Distance to Extinguishers
Light	4-B	50 ft.
Ordinary	8-B	50 ft.
Extra	12-B	50 ft.

For Extinguishers Labeled After June 1, 1969

Type of Hazard	Basic Minimum Extinguisher Rating	Maximum Travel Distance to Extinguishers
Light	5-B	30 ft.
	10-B	50 ft.
Ordinary	10-B	30 ft.
	20-B	50 ft.
Extra	20-B	30 ft.
	40-B	50 ft.

INSPECTING FIRE EXTINGUISHERS

The purpose of an inspection program is to determine whether or not the portable extinguishers are accessible and ready for immediate use in emergency situations.

The first part of an inspection should be directed toward the placement and the accessibility of the extinguisher. The extinguisher should be located in a well-marked area usually adjacent to a door or exit. The location should also be marked with the appropriate sign or symbol. The area surrounding the extinguisher should be readily accessible and free from all obstructions. Extinguishers can only be effective in extinguishing fires when the fire is in its incipient stage. An extinguisher in a well-lighted, highly visible area free from all obstructions will reduce travel time and aid in quick extinguishment.

Inspect for proper locations

A physical examination of the unit is also necessary. The shell should be checked for physical damage such as dents, corrosion or any limitations of its ability to perform effectively. The hose and nozzle should be checked for cracks and obstructions. The lock pin should be in place with the tamper seal attached and unbroken.

The extinguisher should be fully charged. Air pressurized water (APW) and pressurized dry chemical extinguishers are equipped with sight gauges which indicate their charge levels. However, carbon dioxide, some halon, pump tank and cartridge-operated dry-chemical type extinguishers are not equipped with sight gauges.

All extinguishers must be fully charged

The carbon dioxide and halon extinguishers must be weighed to determine if they are fully charged. If a unit is found to be 10 percent deficient in agent by weight, it should be serviced and checked for leaks.

The quantity of agent must be checked in cartridge-operated dry-chemical extinguishers. This can be done by weighing or by removing the cap and checking the agent level. The gas cartridge can be checked by weighing. A 10 percent deficiency in weight of a gas cartridge is due cause for replacement or recharge of the cartridge.

The pump tank should be checked to see if it is full of water. Since the water inside the tank can evaporate, it should not be overlooked. This can be done by simply removing the cap and checking if the water level is at the full line.

After the inspection is complete, the person performing the inspection should record name and the date of the inspection on the inspection tag.

If an inspection reveals that an extinguisher is deficient in its charge, has evidence of physical damage, or raises a question as to its ability to perform properly, it should be removed from service and submitted to a competent person for servicing.

A good inspection program should be carried out at least monthly. Some businesses, depending upon available personnel and the type of hazards, will perform extinguisher inspections as often as once a day.

FIRE DETECTION and ALARM SYSTEMS

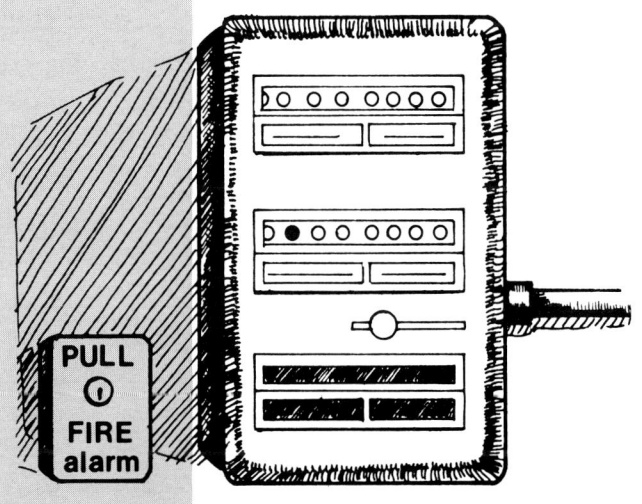

- Types of Systems
- Alarm-Initiating Systems
- Inspection and Testing of Alarm and Detection Systems

FIRE DETECTION AND ALARM SYSTEMS

Signaling systems used for private fire protection purposes are highly technical and include many forms of equipment that are usually installed and maintained by those specialized in this type of work. During inspections or preplanning activities of premises where a system has been installed, fire service personnel should make themselves familiar with the location of annunciators and control panels. Functional aspects of the respective systems should be noted. An observant individual should be able to recognize physical and environmental conditions that may render the system inoperative or unresponsive to a hostile fire. Identification of conditions that may trigger an unwanted alarm can be a positive factor in reducing the number of fire department responses promoted by false alarms.

All of the components in a fire detection and alarm system should be listed by a nationally recognized testing laboratory such as Underwriters' Laboratories, Inc. (UL, Inc.) or approved by the Factory Mutual System (FM) to assure operational reliability. Testing reports may address an entire system or individual components that may be used in interchangeable applications. The installation of the system should also conform to applicable provisions of the NFPA Standard No. 70, *National Electrical Code,* and respective standard for installation maintenance and use of designated types of protective signaling systems as published by the National Fire Protection Association.

Signaling devices conserve lives and property

The expressed purpose of protective signaling systems is to conserve fire losses involving life and property. A relatively simple system will provide a reliable evacuation alarm that can be either manually or automatically activated. Engineering designs may be electronically complex depending on how many functions other than a local evacuation alarm have been incorporated into the system. The size and structural features of the occupancy or protected property, occupancy hazards and age of the equipment will also influence the type of system that may be encountered. A wide variety of new automatic fire detection and signal initiating devices have recently been developed and introduced into the market place. While most of the devices currently in service are designed to produce an electrical signal through an appropriate hard wire circuit, equally reliable wireless equipment has also been perfected. Specific radio and microwave frequencies may be used when conditions warrant to extend protective signaling services into remote or otherwise complex multiplexing capabilities. To remain competitive, most manufacturers also use solid-state electronics and printed circuit boards in their control equipment. These circuits are easily damaged by excessive voltage, reverse polarity and static charges. Also, unless properly protected during installation, maintenance and testing, they can be damaged.

Installation of automatic fire sensors should not be confined to occupied spaces. An appropriate detection device frequently offers the most reliable means of surveillance in uninhabited areas, concealed spaces, or enclosures that are environmentally unsuited for attendant habitation. A typical example involves the use of pneumatically responsive devices in hazardous locations where concentration of air-borne flammable substances may be present. UL listed or FM approved electrical equipment is also available for use in hazardous locations. The current state of the art in detector technology leaves very little to chance and a few places where fire can remain unnoticed, if sound engineering judgment is exercised in preparations of plans for the installation of a system. In the planning stage, an initial decision must be made that establishes an acceptable level of risk. This decision ultimately influences almost every aspect of the system design including ultimate cost.

Where there is a need to automatically protect occupants of a building from the ravages of fire, an alarm system may provide one or more of the following secondary services:

- Shutting down or reversing heating, ventilation and air conditioning systems for smoke control.
- Closing smoke or fire-rated doors and dampers.
- Pressurizing stairwell for evacuation purposes.
- Overriding control of elevators to prevent stops on fire floors.
- Automatically returning elevators to ground level for fire department use.
- Operating smoke and heat vents.
- Activating special fire suppression systems to extinguish or control the spread of fire.
- Notifying the public fire department of an emergency condition.

Alarm systems have secondary functions

At best, there are still many fundamental concerns that have not been entirely resolved through technological advancements in electronics. The ability of an individual to perceive and react to a stressful situation may mean the difference between life and death. To minimize confusion, it is important to remember that audible or visual signals intended to notify building occupants of an emergency condition must have recognizable qualities and should not be used for any other purpose. Applicable evacuation procedures should be conspicuously posted and actively practiced by permanent occupants of the protected property. In compartmentalized high-rise buildings many of the traditional concepts of mass evacuation to a place of safety outside of the building have been changed and now call for restricted movement of endangered occupants to an area of safety above or immediately below the fire floor. If complete integrity is to be maintained and all features of an alarm system are to be effectively utilized, it is incumbent on the fire service to become familiar with this design of the system. Critical features and location of equipment should be noted in the building prefire plan.

LOCAL ALARMS

A local alarm system is a combination of alarm components designed to detect a fire and to transmit an alarm on the immediate premises. The primary purpose of a local alarm system is to alert the occupants and insure their life safety. The secondary purpose of a local alarm system is property conservation. Additionally, local alarms may supervise their own operation to detect the occurrence of system faults, and may operate auxiliary devices such as fire door closers and air-handling shut down equipment. A local alarm system does not retransmit an alarm to any agency or group away from the premises. Installation and maintenance of local alarm systems is addressed in NFPA Standard No. 72A.

The basic components of a local alarm system, and in fact all alarm systems are:

- The control panel
- The alarm initiating devices
- Signaling devices
- Power sources
- Auxiliary devices

These components are shown in Figure 4-1.

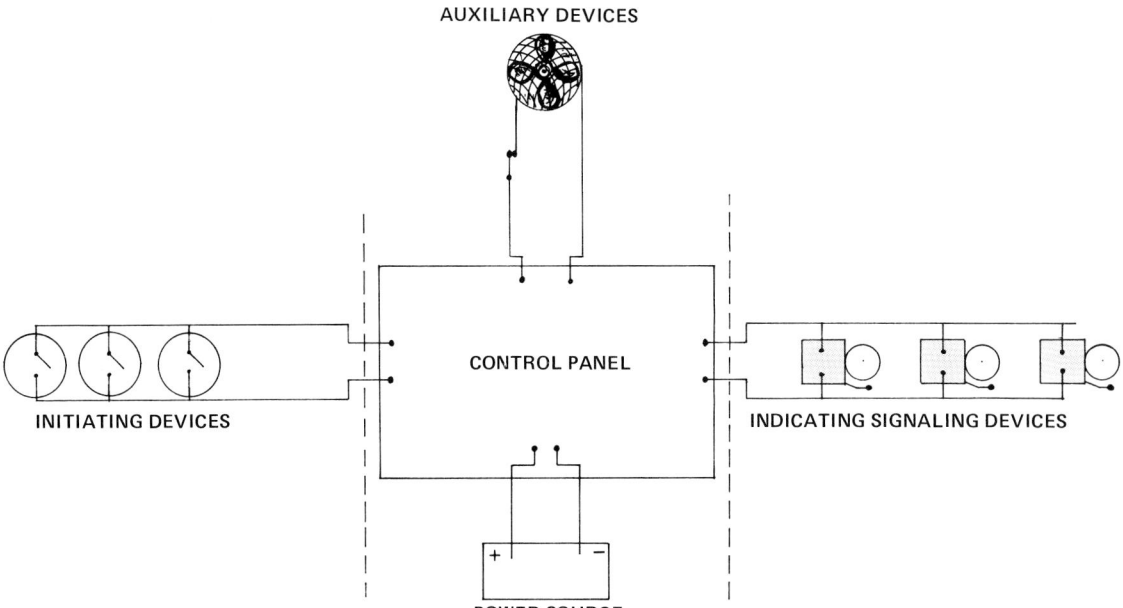

Figure 4-1 The basic components of all alarm systems are the same. The elaborateness or the simplicity depends on the needs of the area to be protected.

The control panel is the central nervous system of the local alarm system and serves many purposes. All the components of the system are connected to the control panel and are controlled from the panel. The hardware contained in the control panel determines the capabilities of the system. These capabilities range from simply receiving a detection signal and sounding an alarm to complicated zoned presignal systems. Figure 4-2 shows a control panel. In addition, the control panel hardware determines at what level the system can self-supervise.

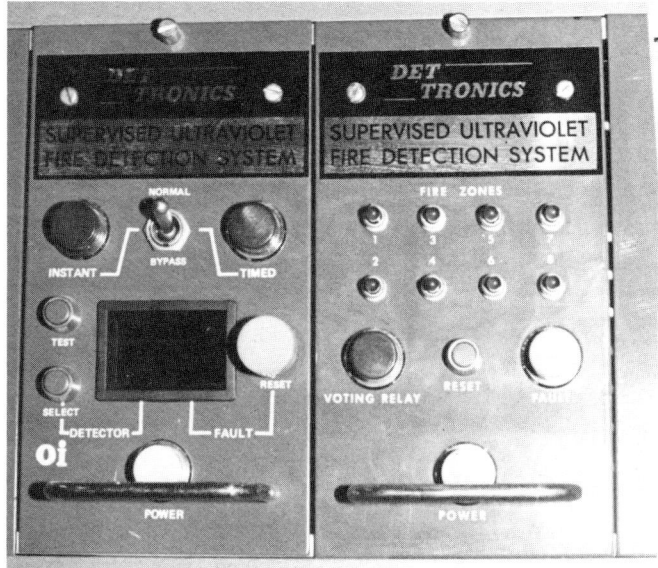

Figure 4-2 The control panel of a system can be very simple or intricate depending on the system. The ultra-violet-detection-system panel is complicated.

The alarm initiating devices are the nerve endings of the system. It is through them that the alarm signal is originated. The basic types of initiating devices are:

- Manual
- Thermal sensitive
- Products of combustion
- Flame detectors
- Extinguisher system related

The alarm-initiating devices monitor the physical conditions in the area around their placement. When fire conditions develop, the device transmits a signal to the control panel which then activates the signaling devices.

Signaling devices are those components which convert the alarm device's initiating action into a form perceivable by the occupants. The most common form is the fire alarm bell, but others include buzzers, chimes, horns, electronic tones amplified through speakers and strobe lights. Standard schematic symbols

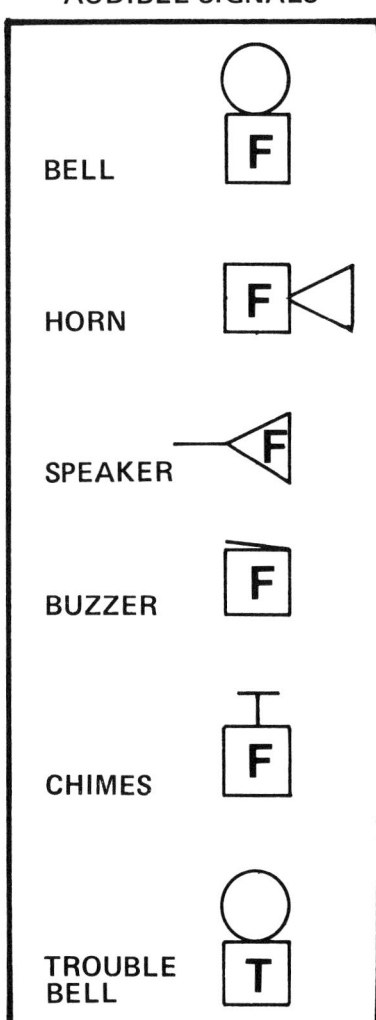

Figure 4-3 The audible-signaling devices can include buzzers, chimes, horns, or electronic tones through speakers. The standard schematic symbols for these are shown.

Figure 4-4 For persons with auditory handicaps special alarm devices such as this visual alarm are necessary.

for audible devices are shown in Figure 4-3. Figure 4-4 shows a strobe signaling device. The alarm signal can be noncoded, zone noncoded, master coded or zone coded. Each type can be of a general or presignal design.

A noncoded system is the simplest type of system. All incoming signals from alarm-initiating appliances connect to the same alarm-panel-initiating circuitry. In other words, every alarm-initiating device will sound an identical alarm signal. Usually, these are continuously sounding alarms. The panel has no capability to differentiate between which alarm-initiating device sent the alarm or to produce more than a continuous alarm signal. Figure 4-5 shows a typical noncoded system.

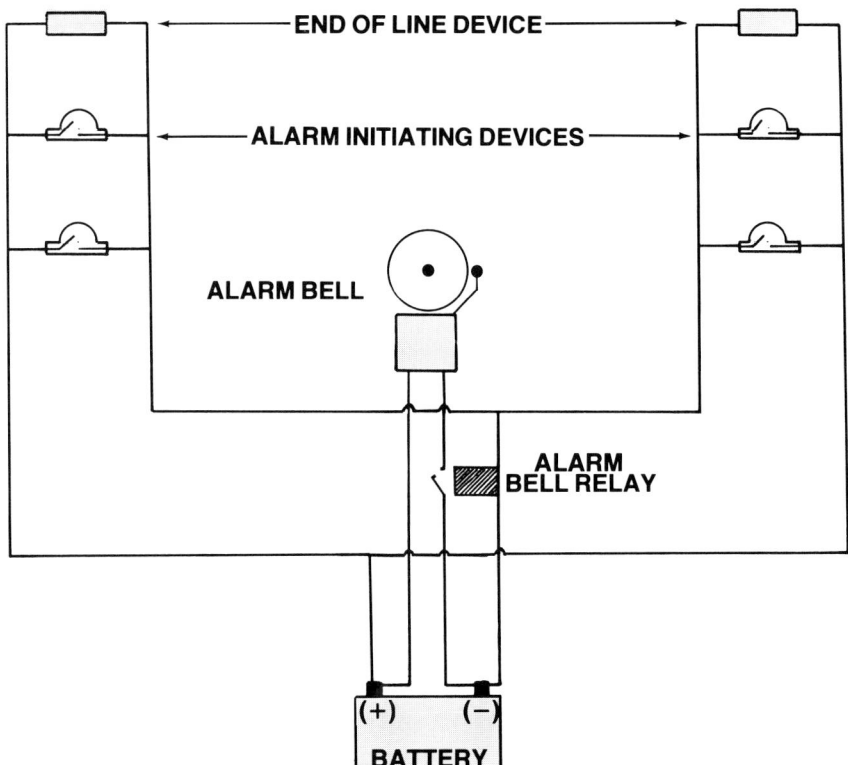

Figure 4-5 A noncoded system has no capability to differentiate which alarm-initiating device sent the alarm, or to produce more than a continuous signal.

A zone noncoded system is the first progression up from a noncoded system. The difference, as with most improvements in local systems, is the alarm panel circuits. In a zone noncoded alarm system, the alarm-initiating devices are received into the panel in groups from common areas. Each zone is monitored by an indicator lamp placed in the circuit. The small flow through the supervisory circuit is not strong enough to light the lamp. When one of the initiating devices closes, causing the current to flow, the current lights the lamp.

The alarm bell is maintained in a separate circuit controlled by an electrical relay. The relay is placed in the circuit with the alarm system. However, like the light, the small supervisory circuit is too weak to close the contacts. When an initiating device switches on, the relay opens and closes, sending intermittent current to the alarm bell. The zone noncoded panel still will not produce anything but one alarm signal. This type of system allows the responding emergency personnel to determine the zone from which the alarm was initiated by visually checking the alarm panel to see which lamp is lighted. Figure 4-6 shows a zone noncoded system.

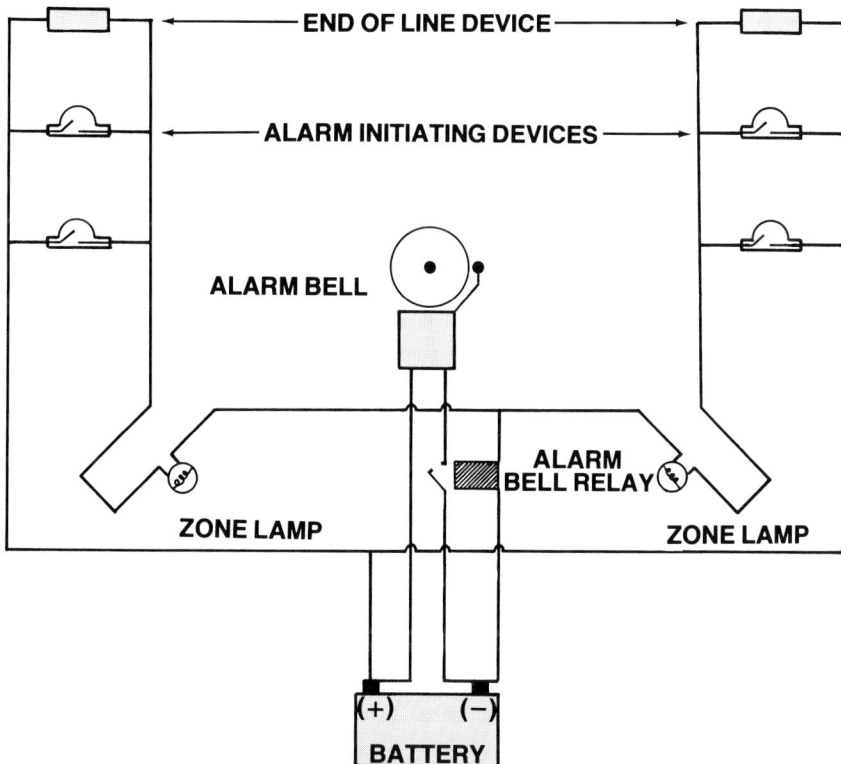

Figure 4-6 A zone noncoded enables the responding personnel to determine in which zone the alarm initiated, but has only the capability to produce a continuous alarm signal.

Master-coded systems are systems which are equipped with a panel which generates a code through the signal devices. Generally, this is a series of short alarm signals instead of one continuous signal. It could be simply a relay which opens and closes, causing an intermittent alarm signal. On complicated alarm systems, sequences might be employed such as a predetermined number of short rings followed by a predetermined number of long rings. However, the panel is only capable of producing one alarm signal. Every alarm will sound the same regardless of the location of the original alarm-initiating device.

Master coded systems are employed in occupancies that use the alarm signals for other purposes. For example, a school using the same bells for sounding class changes as it does for fire alarms. Obviously, this could cause confusion. By putting a master coder on the system, the fire alarm would have a distinct and recognizable sound. It is best to use a master code which starts with a short ring cycle. Often a coder can sound five or six short rings in the time a normal class-change ring would last. In this instance, it would be hard to mistake a fire alarm for a class change. Figure 4-7 shows a schematic of a master-coder system.

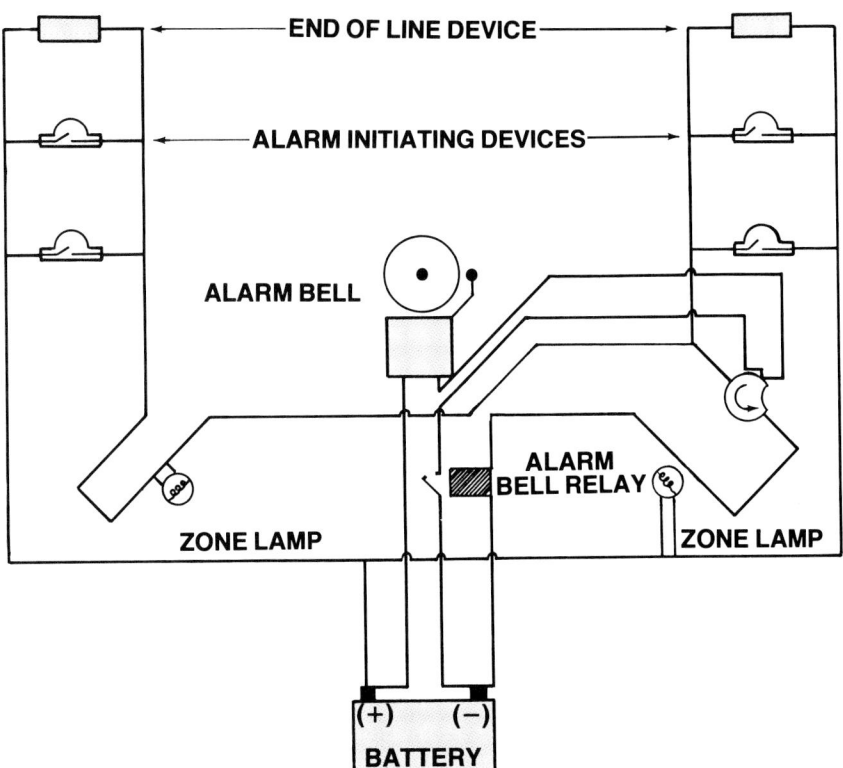

Figure 4-7 Master-coded systems are utilized in schools and other occupancies where bells are in normal usage. The master coder produces a distinct and recognizable alarm signal over the bell system.

Zone-coded systems resemble zone noncoded systems, except instead of, or in addition to, adding a lamp to monitor the zone circuit, a signal-coding device is placed into the circuit. This signal device will sound in a specific and unique pattern for each zone. Systems utilizing the zone-coded arrangement enable employees on the fire brigade to determine the zone of the emergency simply by listening to the pattern of the signal.

Usually, the pattern is a series of short rings, then a brief pause followed by a second series of short rings. A long pause will follow the first set of rings. Then the cycle will repeat. Most systems are designed to code the first zone which comes in and to disregard subsequent alarm signals. A zone-coded alarm schematic is pictured in Figure 4-8.

Coded signals in local systems are sometimes confused with telegraphic fire department alarm boxes that are called selective coded systems. There two groups of devices are not the same. Local coded systems produce a coded audible or visible alarm as a

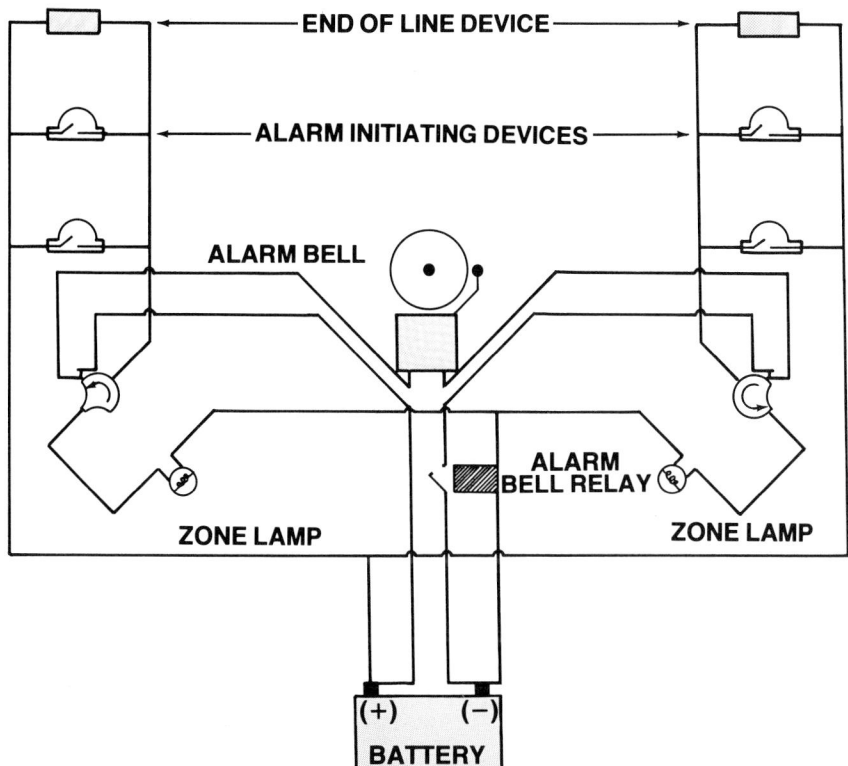

Figure 4-8 A zone-coded system can produce a separate audible signal for each zone on the system. This allows emergency personnel to determine the zone of origin.

result of electronic circuitry located in the alarm panel. The alarm-initiating device only transmits the presence of an open or closed circuit. The location of the attachment of the particular alarm-initiating device into the alarm panel and the subsequent alarm panel circuit determine the code enunciation. Selective coded systems, on the other hand, transmit a coded signal to the alarm receiving station. They accomplish this through the use of some type of coder *incorporated into the alarm-initiating device.*

There are three basic devices which provide coded signals. The oldest method is the eccentric wheel. The wheel is mounted on a drive motor which is either spring- or electronically powered (Figure 4-9). A set of flexible contacts rest on the wheel. The

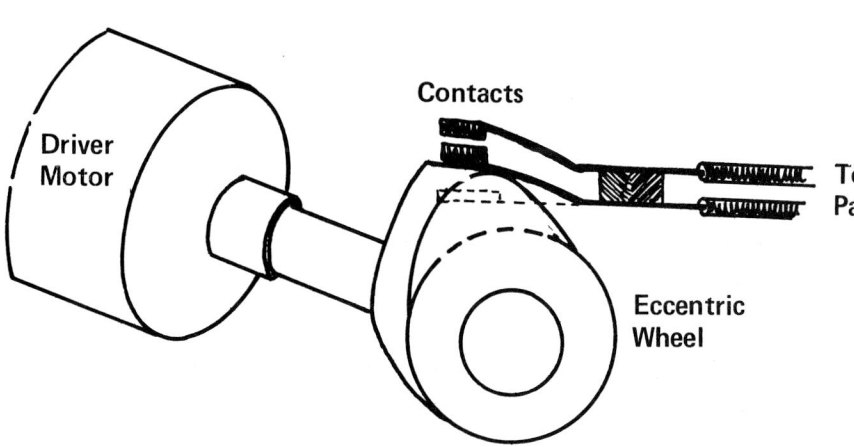

Figure 4-9 An eccentric or notched wheel can be used to open or close electrical contacts sending a coded signal.

motion of the wheel results in the contacts opening and closing, causing a coded signal to be sent. The second method uses electronic relay circuits. These relays are electronically interconnected so that a current flowing through the circuit causes them to open and close in a predetermined pattern. The simplest of these relays is similar to the blinker on an automobile. The final method or device is the solid-state circuit board. These devices are electrical, printed circuits which can be used effectively to produce a distinct code pattern.

In addition to sounding an alarm, a local system can activate or deactivate other auxiliary devices. These devices include fire doors and dampers, lights, elevators, air movement equipment and fire control systems. These auxiliary functions can be activated by zone or in a general mode.

A presignal must be necessary before the alarm

A presignal alarm is employed in an occupancy in which the potential for panic is high, such as in a hospital. The system responds initially with a presignal to alert emergency personnel prior to the general occupancy being notified. This presignal is usually a discreet signal recognizable only by persons familiar with the system. Sometimes this is a prerecorded message over an intercom, or it could even be a pager signal. This provides the emergency personnel an opportunity to assist the general occupancy in exiting. The emergency personnel may elect to handle the situation without sounding a general alarm, may sound the general alarm after investigating the situation, or may allow the alarm panel, if so designed, to initiate a general alarm after a predetermined time delay.

PROPRIETARY SYSTEMS

Local alarm systems can be thought of as the basic alarm setup upon which other alarm systems can be built. The basic system involves coverage of a single occupancy. The alarm is not monitored except when the alarm signal is sounded and someone hearing the alarm responds. The alarm system does not register the existence of an alarm away from the building of origin.

In large industrial plants in which several buildings are protected, a central alarm receiving point is developed. This central point has a console which is constantly manned by competent personnel. These persons have knowledge concerning alarm systems, system operation and system maintenance in addition to being trained to handle emergencies. These systems are known as proprietary systems. A graphic representation of a proprietary system is shown in Figure 4-10.

The systems can be zoned or coded similar to a local system. Each building may have its own code, or a certain block or area may have a specific code. Usually, however, the building location is discernible by annunciator panel lamps.

Upon receiving the alarm, the personnel manning the central alarm point take the appropriate action. The action may include notifying the fire department or sending a service crew to investigate a trouble signal. For specifications concerning proprietary systems, see NFPA Standard No. 72D, *Proprietary Protective Signaling Systems.*

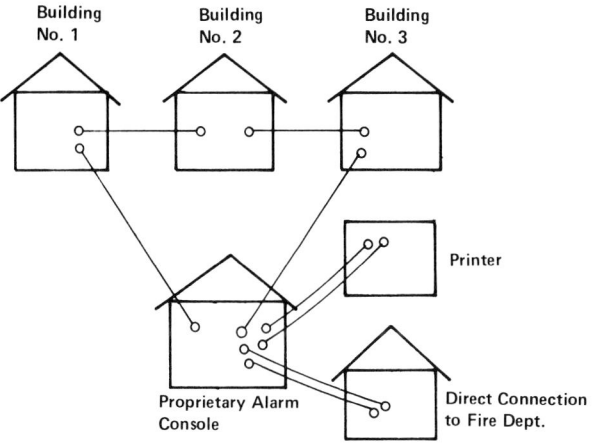

Figure 4-10 A proprietary system involves a user-owned control center for monitoring several separate alarm systems on the same property.

CENTRAL STATION

A proprietary ties several separate local alarm systems into a central location owned and operated by the owner of the property. Often, several small industrial plants with one or two buildings each will be protected under one system since they are not large enough for the maintenance of manned alarm centers to be economically feasible. In these instances, a separate contractor may choose to develop his own alarm receiving station, then contract to different plants to provide monitor service. The service provided and the requirements for manning the station are essentially the same as found for a proprietary system. The difference lies in the fact that a central station receives alarms from different properties not under its ownership. A central station layout is shown in Figure 4-11.

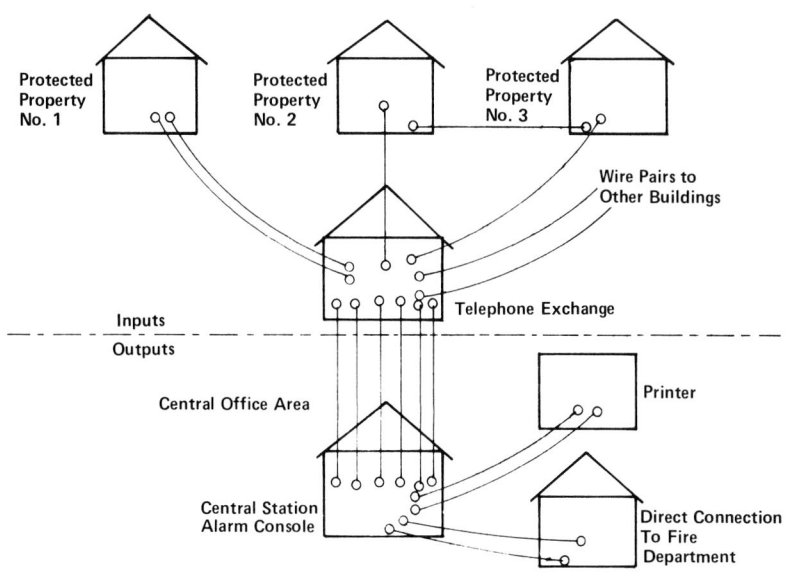

Figure 4-11 A security firm secures contracts with several occupancies to provide a manned alarm center. Usually installation service and maintenance contracts are available from the central station service.

Most of the time, the alarm is received at the central station via leased telephone equipment; however, some central station services may provide their own transmission systems. For specifications on a central station system, see NFPA Standard No. 71, *Central Station Signaling Systems*.

REMOTE STATION

In areas where the private contracting of central station service has not been established or maintained, the community dispatch service or the local fire station may be equipped to receive alarm signals from local alarm systems. This provides a direct-dispatch capability. Supervision of the systems for maintenance and breakdowns is often a problem. The municipal service is usually unable to provide any maintenance service other than notifying the owner that a trouble signal has been received. In smaller towns or in rural counties, the local law enforcement dispatcher's office is often chosen as the location of the remote station. This is because the fire station or headquarters is not constantly manned. Often, this has resulted in less than desirable results. These conditions are the result of the personnel manning these locations not being trained or attuned to fire protection requirements. A remote station system is shown in Figures 4-12 and 4-13.

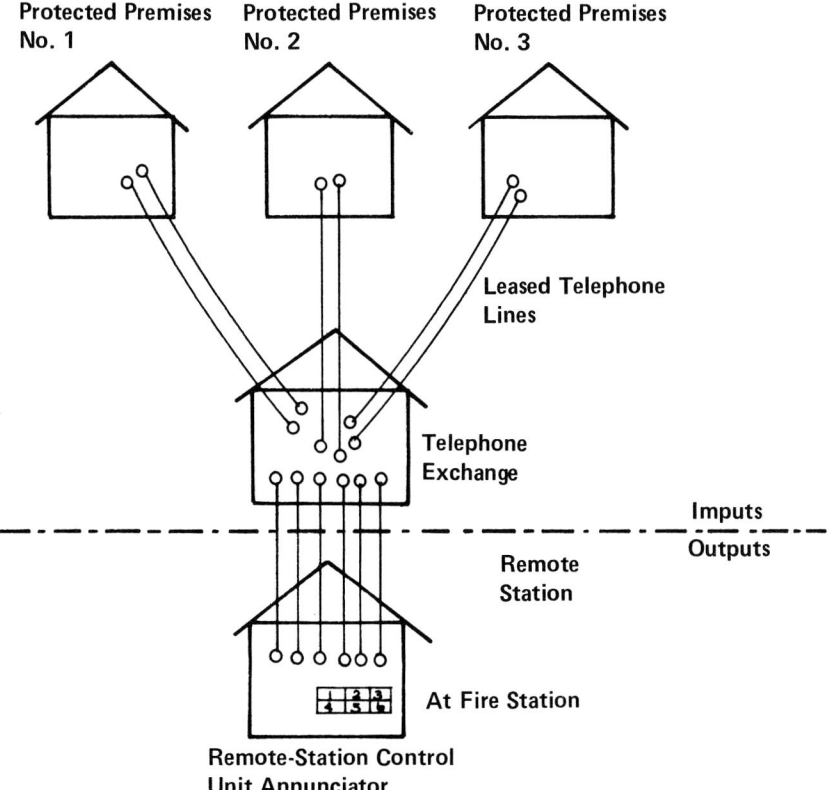

Figure 4-12 A remote system ties into a municipal dispatch facility.

Figure 4-13 The remote station alarm annunciator panel located in a police dispatch center.

MUNICIPAL FIRE ALARM SYSTEMS

Although the main purpose of municipal fire alarm systems is to allow the public to notify the fire department of an incident, they can and are used with private fire protection systems. The private alarm system, either a local or proprietary system, can be interconnected to the municipal fire alarm system. The private system is then classed as an auxiliary system.

Municipal systems are either Type A, Manual or Type B, Automatic. The signal transmitted through a Type A system to the fire department alarm office has to be retransmitted over alarm circuits to fire stations. A Type B system automatically retransmits to all fire stations signals received from all system alarm boxes. A private auxiliary system would be connected to the municipal system through a master box and would transmit the signal to the fire department alarm office.

There are three types of auxiliary systems in use: local-energy type, shunt-type and direct circuit. The local-energy and shunt-type systems are used in telegraph systems. A direct-circuit system would be used with a telephone or radio master box.

A local-energy auxiliary system receives its power supply from a source other than the municipal system. An initiated signal is relayed to the municipal circuitry for transmission to the alarm office.

A shunt-type auxiliary system is directly connected to the municipal alarm circuits. On this type of system, only water flow and manual pull stations can be used, and they become in essence a part of the municipal circuit.

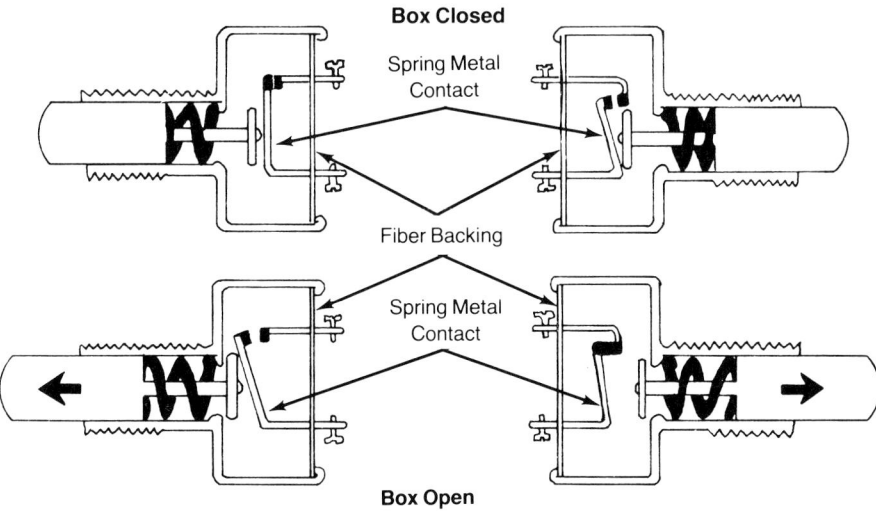

Figure 4-14 An electrical switch is utilized in the typical pull box. Some are designed to make contact while others are designed to break contact when the box is opened.

Figure 4-15 Manual activation stations, commonly called pull boxes, come in a variety of sizes and shapes.

 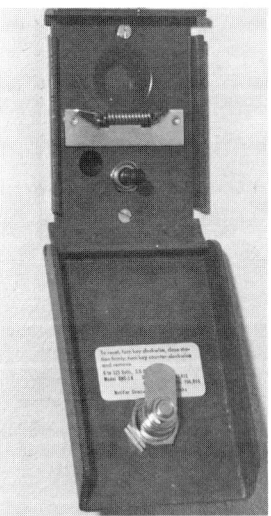

Figure 4-16 A pull box is an electrical switch set in a frame in a manner than when the box is pulled the switch opens or closes.

A direct-circuit auxiliary system goes from the protected property to the fire department alarm office by a direct individual circuit.

The private system is connected to the municipal alarm circuit and initiates a signal just as if a citizen were to use the public manual station. The circuit from the master box is supervised as the other municipal direct circuits.

Master box circuits are supervised

An accidental opening of the alarm circuit on a shunt-type system will cause an alarm to be transmitted. Also, a ground on the alarm circuit of a shunt or direct-wire auxiliary system could result in an impaired transmission. Many boxes will have to be rewound and/or reset after use. A policy should be established as to who will perform this duty after alarms or tests.

The use of auxiliary systems may become less as cities remove their municipal alarm systems. Difficulties can develop between the private system owner and the city if there are excessive false alarms. This may lead to the removal of manual alarm systems which are accessible to the public. New technology in the electronics and communications fields should serve to improve the reliability on any new or updated auxiliary systems.

NFPA Standard No. 73, *Public Fire Service Communications* and Standard No. 72B, *Auxiliary Signaling Systems* can be referenced for further information on installation and maintenance of these systems.

ALARM INITIATING SYSTEMS

Any alarm system, whether it be a small local system or a major multiplex high-rise system, depends on the same basic alarm-initiating devices. The devices are the sensors for the alarm system. The system depends exclusively on them to detect fire conditions. There are three major types of alarm-initiating devices: manually activated devices, products of combustion detectors and extinguishing-system activation devices.

Alarm-initiating devices are the sensors of the alarm system

Manually Activated Devices

Manually activated devices mostly commonly are pull boxes. These devices consist of a set of electrical contacts on a manual switch. The switch is preset in a fashion where the electrical contacts are either open or closed. When someone manually moves the switch, the contacts are moved (Figure 4-14). The opening or closing of the contacts changes the electrical current's status, which is monitored by the alarm panel. This change is arranged to activate an alarm signal. Manual activation stations come in a wide variety of shapes, sizes and designs. Figure 4-15 shows typical pull boxes. Their operation is similar regardless of shape or size (Figure 4-16).

Figure 4-17 A pull box with a glass cover that must be broken for use is not recommended. The breaking tool is often removed and the broken glass presents a safety hazard to the user.

The only type of manual activation station which is not recommended is the type in which a glass cover must be broken out to get to the activation switch (Figure 4-17). These types of boxes were originally designed to go into areas where the misuse of the activation switch was a problem. Some of these stations were accompanied by a small striking tool to assist in breaking out the glass. These devices have somewhat reduced the amount of false alarms, but they constitute a safety hazard to the user because of the glass breakage. Some such devices have had the glass replaced with plastic covers which can be removed or broken easily. Figures 4-18 and 4-19 show two additional approved pull boxes.

Figure 4-18 A pull box that requires the breaking of a glass rod is acceptable. The breaking does not require a special tool and does not cause a major hazard to the user.

Figure 4-19 In some industrial and institutional occupancies master boxes are utilized as pull stations.

Products-of-Combustion Detectors

Products-of-combustion detectors include the group devices generally called fire detectors. Their operation is designed to occur when one of the four major groups of detectable products of combustion occur. Figure 4-20 shows the detectable products of combustion. The four major detectable products of combustion are heat, visible products (smoke), invisible products (ionized gases) and light. Fire has other products of combustion, but those listed are the general groups which, in addition to being detectable, are indicative of fire. This is to say that the amount, concentration, or degree of the product is discernible from other naturally occurring products. Therefore, when the product is detected, it is acceptable to assume the existence of a fire is at least highly probable. An example of a product of combustion which would not be distinguishable would be carbon dioxide (CO_2). The presence of

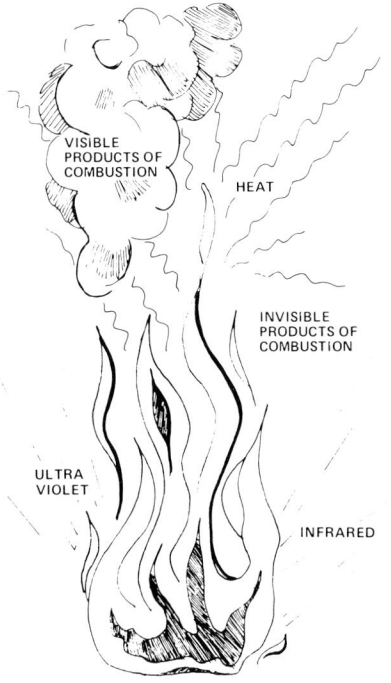

Figure 4-20 Fire produces four easily detectable products of combustion. These are heat, invisible products in the form of ionized gases, visible products in the form of smoke, and light. The ultra-violet and infrared spectrums of lights are most commonly used for flame detection.

CO_2 is easily detectable by some devices, and it is a product of combustion in a high percentage of all fires. The problem is that carbon dioxide is the fourth most abundant gas in the atmosphere, and it is a by-product of animal and plant respiration and many other processes. Therefore, its presence could not be indicative of the presence of fire.

HEAT DETECTORS

Heat is an abundant product of combustion. It is detectable by certain devices using three primary principles of heat physics. First, heat causes the expansion action of material; second, it causes a melting action; and third, the thermoelectrical characteristics of heated metal are detectable. Three groups of devices use these primary principles to detect fire. These are the fixed temperature, rate-of-rise and combination fixed-temperature rate-of-rise devices.

1. FIXED TEMPERATURE

The fixed-temperature devices are designed to activate at a given temperature. The two most common types use the characteristic of melting material to detect fire.

- The first type of fixed temperature devices is bimetallic, which use the expansion characteristics of heat. Two metals or metal alloys with different expansion ratios when heated are used. Those metals are formed into thin strips and bonded together to form a single strip. The effect of heat on the combined metal (bimetal) strip will make the metal with the larger expansion ratio expand more rapidly. This will cause the bimetal strip to arc toward the side of the strip made up of the metal with the lower expansion ratio. The amount of curve and the difference in expansion of the two metals at a specific temperature is mathematically or experimentally calculated. Then the bimetal strip is placed into a housing in a manner that causes two electrical contacts to close when a given amount of arc is reached.

The bimetallic strip is most often placed with only one end held secure and the other allowed to move. In some devices, however, both ends of the strip are secured with a slight bowing toward one side. When heated, one side of the strip compresses because of the expansion of the faster-expanding metal. Finally, when the expansion is great enough, the strip snaps to bow in the opposite direction. This action opens or closes a set of electrical contacts, which initiates a signal. (Figure 4-21 shows a schematic of this type detector.)

Most bimetallic detectors are considerd to be self-resetting. This gives them the advantage of not having to be replaced after a fire. It is important, however, to check the detectors after a fire to insure that permanent damage has not resulted.

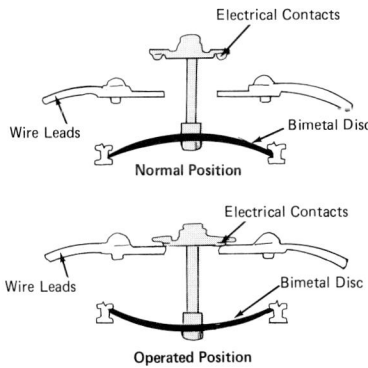

SNAP ACTION DISC THERMOSTAT

Figure 4-21 A bimetallic strip housed in a bowed position can be designed to invert when heated. This principle is used to manufacture restorable fixed-temperature detectors.

- The second type of fixed-temperature detection devices are based upon the principle that most materials will melt when exposed to heat. Moreover, the melting temperature of most metals is very specific. This is to say that a solid, e.g., ice, melts at 32°F, and the temperature for this melting will not change. Therefore, if a device was needed to send an alarm when the temperature reached 32°F, ice could be used by placing a small layer of ice between two contacts. When the ice melts at 32°F, the contacts will come together sending an alarm.

Melting solder parts can be used to initiate an alarm signal

Realistically, soft metal alloys and thermoplastic resins are used for these applications. Solder is often utilized for this purpose. By altering the alloy composition of a metal, a very specific melting temperature can be achieved. The solder is usually utilized to hold a two-piece link or latching mechanism together. When the solder melts, the link separates or the latch is released, causing an alarm signal to be sent.

Plastic insulated wires will send a signal when the plastic melts causing a short

Thermoplastic polymens can also be manufactured to melt at a very specific temperature. These polymens are used to insulate steel wires which are then wound around each other to form a tension cable. When the melting temperature of the polymen is reached, the tension of the wires causes the wires to twist through the plastic and short together, sending an alarm. These cables can be run along large areas forming a continuous fixed-temperature alarm device. After a fire, the melted section of cable is cut out and replaced with new unmelted cable.

Another type of detection cable, which employs the melting of a chemical salt insulation, is utilized. The chemical salt is not electrically conductive as a solid, but it is very much so when a liquid. The temperature at which the salt melts can be carefully controlled by adding certain chemicals. The cable can be designed to signal when the temperature reaches a predetermined level. The advantage of this type of cable over the thermoplastic is that it does not have to be replaced after a fire, because as soon as the salt resolidifies it no longer conducts current. The disadvantage is that it is more limited in its application capabilities since sharp bends, twisting and connections are more difficult to achieve.

Cable detectors are normally referred to as line-type detectors while noncable are known as spot type. Figure 4-22 shows both thermoplastic and chemical-salt line-type detectors.

- The next type of fixed-temperature devices utilize the expansion of heated solvents. A solvent, like a liquid, expands and vaporizes when heated. This means the vapor pressure increases. The solvent is placed into a small glass bulb. The bulbs are commonly called frangible bulbs and are uniformly manufactured to break at predetermined pressures. By

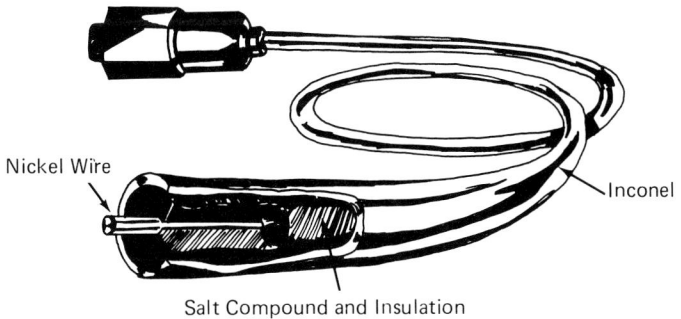

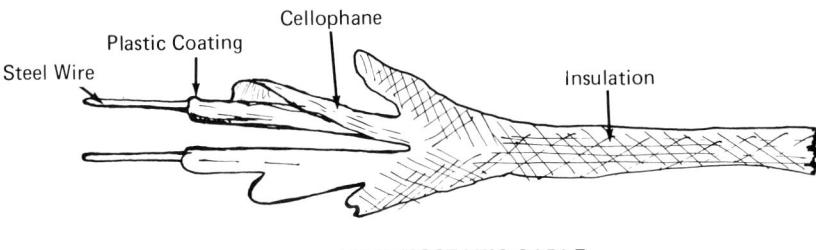

THERMOSTATIC CABLE

Figure 4-22 Thermoplastic and chemical salt fixed-temperature detectors provide an excellent source of linear-type fire detection.

matching a solvent's vapor pressure at a given temperature with the designed breaking strength of the glass bulb, the bulb can be designed to break at a given temperature. The bulb is then placed into a detection device in a manner which separates two contacts. When the bulb breaks, the contacts close, sending a signal.

Fixed-temperature devices are considered extremely reliable detectors, but their sensitivity is considered very low. Most detectors of this type must be replaced after a fire. Since they are relatively simple detectors, their cost is minimal; and for this reason, they are found on many systems. Figure 4-23 shows temperature ratings of detectors.

DETECTOR RATINGS

Normal Ceiling Temperature		Temperature Rating of Detectors
Ordinary	Less than 100° F.	135° to 160° F.
Intermediate	100° to 150° F.	175° to 225° F.
High	150° to 225° F.	250° to 300° F.
Extra High	225° to 300° F.	325° to 360° F.

Figure 4-23 Detector ratings are determined by the temperature to be encountered.

2. RATE-OF-RISE DETECTORS

Rate-of-rise detectors depend almost exclusively on the expansion characteristic of heat. The main exception is the thermoelectric rate-of-rise detector.

Most rate-of-rise detectors employ a small chamber filled with air, with the bottom made of a flexible metal diaphragm. These are known as pneumatic rate-of-rise detectors. As the air inside the chamber expands, the diaphragm is forced out. When it is forced out to a predetermined level, it forces a set of electrical contacts to open or close. This change of current initiates an alarm signal to the alarm panel.

- The most common pneumatic rate-of-rise detectors are individual detectors, known as spot detectors. These detectors appear to be semispheres mounted on a base. Figure 4-24 shows a detector of this type. These are typically brass colored. It must be noted, though, that if they have been painted by other than the manufacturer, they must be replaced. The painting of any detector changes its detection capabilities.

In addition to the diaphragm described earlier, the rate-of-rise detector is vented to avoid a change in ambient temperatures or barometric pressure setting off the alarm. The vent has to be of a size that will allow air to leave or enter the chamber at a predetermined rate below the rate which is necessary to set off the detector.

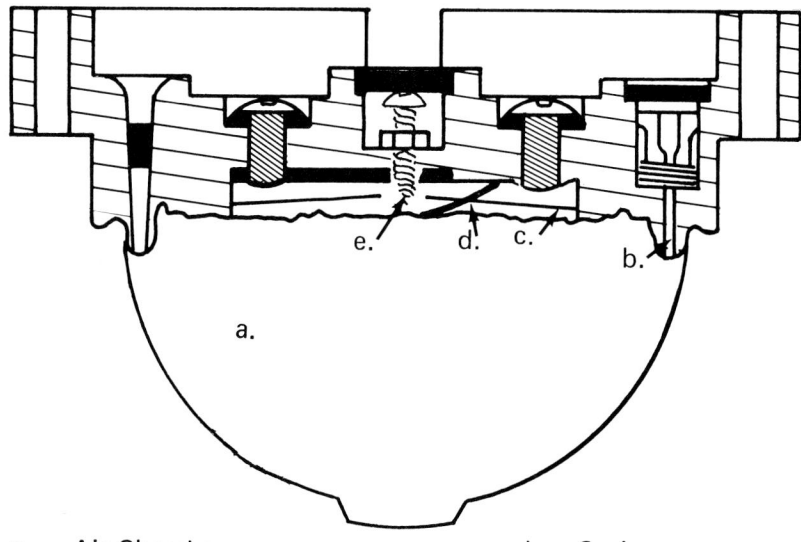

a. Air Chamber
b. Vent
c. Flexible Metal Diaphram
d. Spring
e. Adjustment Screw

RATE OF RISE TEMPERATURE DETECTOR

Figure 4-24 A pneumatic rate-of-rise detector utilizes a pocket of expanding air to detect fire. The pocket is vented to allow the detector to adjust to moderate temperature and barometric changes.

Fire Detection and Alarm Systems **145**

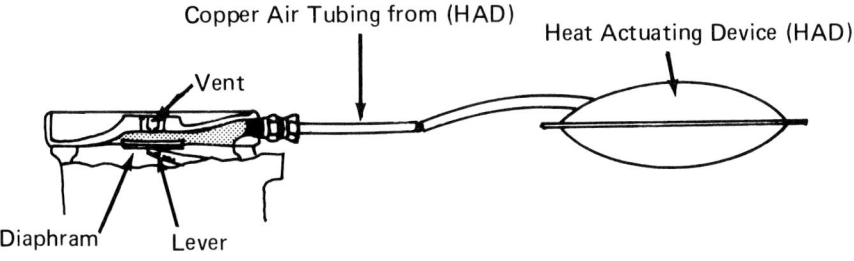

PNEUMATIC CONTROL IN SET POSITION

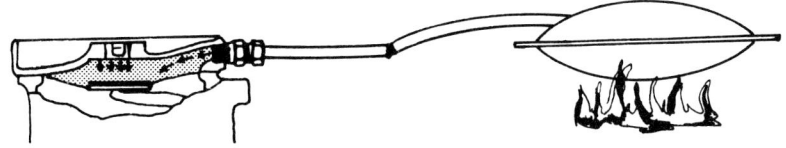

PNEUMATIC CONTROL IN RELEASE POSITION

Figure 4-25 A pneumatic-tube system can be equipped with large air chambers which increase the sensitivity of the line-type detection system. These devices are usually called HADS which stands for Heat Activated Device. *(Courtesy of Walter Kidde & Co., Inc.)*

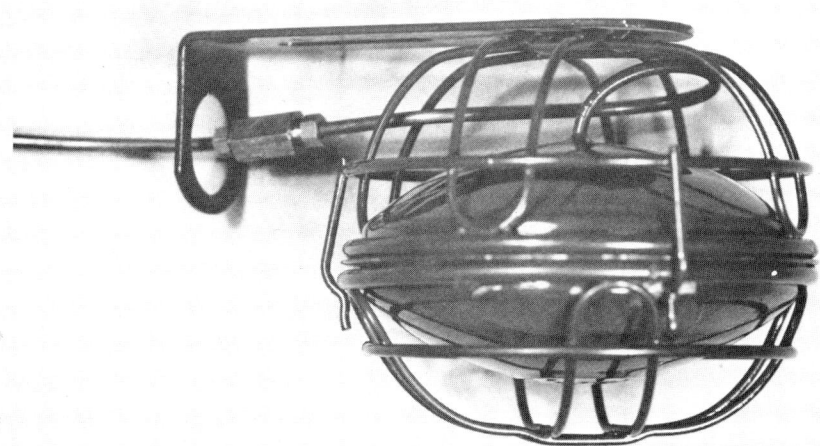

Figure 4-26 A HAD is sometimes enclosed in a wire protector to prevent physical damage to the detector.

- A second type of pneumatic rate-of-rise temperature detection system uses tubing arranged over the area of coverage. The space inside the tubing acts as the chamber. The tubing is attached to an activation vessel that has a flexible diaphragm. This diaphragm performs the same function as those in the base of the spot detector's chamber and is vented similarly. Some pneumatic-tube detection systems may have large vessels or heat activation devices piped in to monitor a specific area of coverage. Figure 4-25 shows the workings of this application. This increases the volume over a target area and adds to the sensitivity of the system at that particular area. Figure 4-26 shows a typical heat activated device.

- Another type of rate-of-rise detector is known as a rate-compensated detector. This detector uses a bimetallic activation principle. A metallic sleeve with a moderate expansion rate, holding two bowed struts of a slow expansion rate, is used to accomplish the necessary rate adjustment. The lower-expansion-rate strips are bowed and held secure at both ends inside the metal sleeve. The strips have electrical contacts attached. In the normal position, the strips bow out and do not allow the contacts to come together. As the temperature increases rapidly, the metal sleeve which houses the strips elongates more rapidly than the strips. This action relieves the tension on the strips and allows the contacts to come together. If the rate of temperature change is relatively slow, less than 5°F the metal strips expand at a rate fast enough to maintain the tension inside the sleeve. This prevents the contacts from coming together. Figure 4-27 shows the workings of a rate-compensated detector.

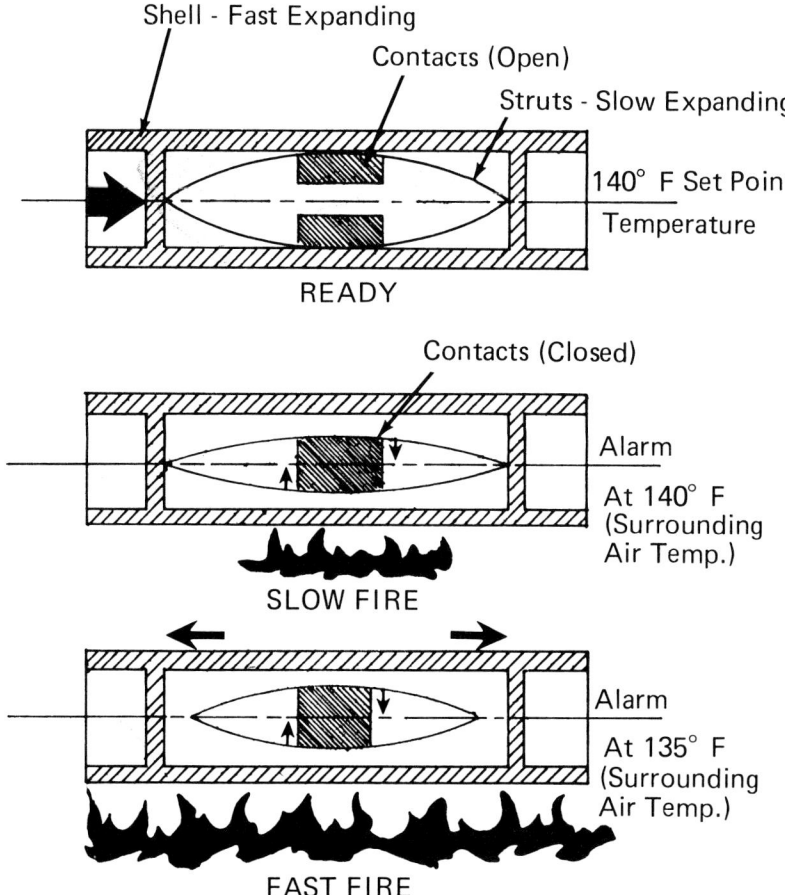

Figure 4-27 A heat-compensated detector uses the expansion characteristics of two different metals to produce one of the most reliable rate-of-rise detectors.

- The last type of rate-of-rise detector utilizes thermoelectric sensors to detect rapid changes in temperature. The principles of a thermoelectric sensor involves the fact that two wires of dissimilar metal, when twisted together and heated at one end, will cause an electrical potential to be generated at the other end. This potential is detectable as electrical current. The rate at which the thermoelectric element is heated will determine the rate at which the current is generated. The detector is electronically designed to bleed off or dissipate small currents. This allows it to disregard small or gradual temperature changes. However, when the change is greater, the increase in currents transmits an electrical signal to the alarm panel. Figure 4-28 shows a thermoelectric rate of rise detector.

Rate-of-rise detectors tend to be more sensitive (react sooner) than fixed temperature. Unfortunately, the fact that these detectors are more complicated makes them slightly less dependable than the fixed-temperature. Except for the rate-compensated, which is considered as dependable as fixed-temperature detectors, rate-of-rise tend to be more expensive than fixed temperature.

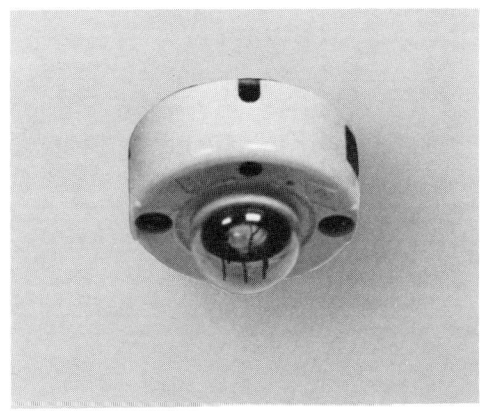

Figure 4-28 A thermoelectric rate-of-rise detector uses the electricity produced by heated dissimilar metals that have been twisted together to detect fire.

DETECTOR	SENSITIVITY	RELIABILITY	STABILITY
Fixed Temperature	Poor	Good	Good
Rate-of-Rise	Fair	Fair	Good
Rate Compensated	Fair	Good	Good
Invisible Products-of-Combustion	Good	Fair	Fair
Visible Products-of-Combustion	Good	Fair	Fair
Flame - Ultraviolet	Good	Fair	Fair
Flame - Infrared	Good	Fair	Poor

Figure 4-29 Each detector has its own good points and bad points. The more complicated the detector the more sensitive. Usually though, this makes the detector less reliable and stable.

3. COMBINATION RATE-OF-RISE FIXED-TEMPERATURE

The last group of heat detectors to be discussed is the combination rate-of-rise fixed-temperature. These detectors are exactly what the name implies; a combination of a rate-of-rise detector with a fixed-temperature. This allows the higher sensitivity of the rate-of-rise to be coupled with the higher dependability of the fixed-temperature detector (Figure 4-29). Figures 4-30 shows a typical combination rate-of-rise temperature detector.

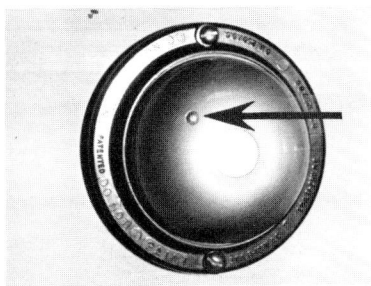

Figure 4-30 The dot on the outer shell should be visually checked to ascertain if the fixed-temperature spring has been released. The small nipple will be missing and a hole will be noticed if the spring has been released.

A spring tension strip is soldered into a standard rate-of-rise detector to produce a combination rate-of-rise fixed-temperature detector. When the solder melts the strip springs against the diaphragm sending an alarm. Refer to Figure 4-24. *(Art courtesy of Pyrotronics, a subsidiary of Baker Industries)*

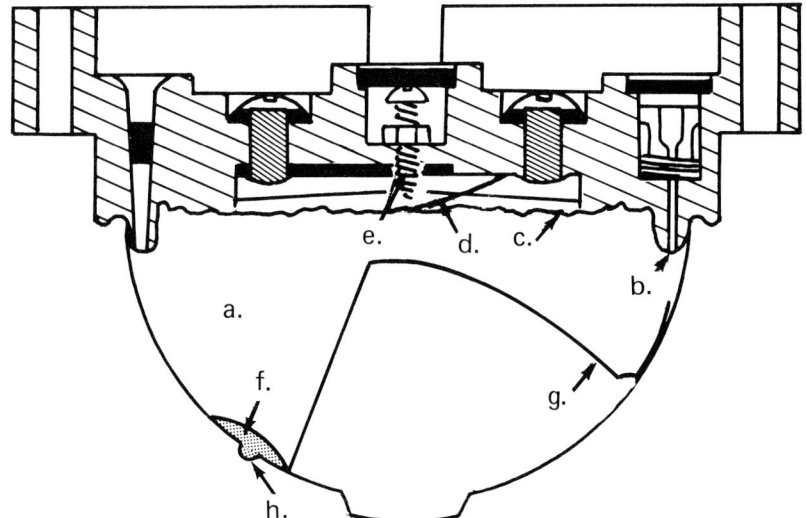

a. Air Chamber
b. Vent
c. Flexible Metal Diaphram
d. Spring
e. Adjustment Screw
f. Fusible Alloy
g. Spring
h. Indicator

COMBINATION RATE-OF-RISE AND FIXED-TEMPERATURE DETECTOR

VISIBLE-PRODUCTS-OF-COMBUSTION DETECTORS

Visible-products-of-combustion smoke detectors are manufactured utilizing a photoelectric cell coupled with a specific light source. The photoelectric cell is a flat disc which when exposed to light transforms the light into electrical current. This cell is employed in two fashions to detect smoke: the first is the beam application; and the second is the refractory application.

The beam application utilizes a beam of light focused across the area being monitored onto the photoelectric cell. The cell is constantly converting the beam into current. This current is used to electronically keep a switch open. As long as the beam is striking the photoelectric cell the switch is kept open. When smoke blocks the path of the beam the current ceases to be produced and the switch falls closed, initiating an alarm signal. Figures 4-31 and 4-32 show line-type photocell applications.

The beam can be focused across any area to be monitored. This can be a large room or simply across an air outlet. A single beam can be reflected by use of mirrors or prisms to extend coverage. When the beam application is used a delay is generally programmed into the alarm switch. This delay will allow an animal or the top of a piece of moving equipment to move through the beam without sounding an alarm.

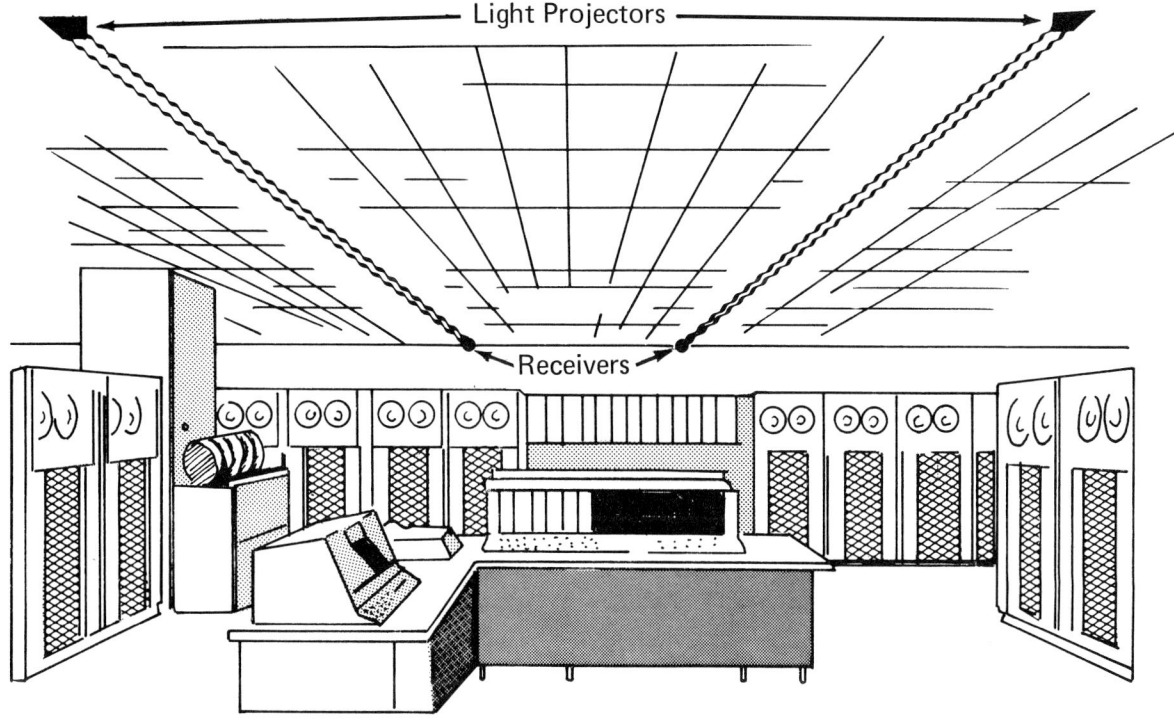

LINE—TYPE SMOKE DETECTOR

Figure 4-31 A line-type photocell detector is capable of monitoring large open areas. The atmosphere must be kept relatively clean to avoid false alarms.

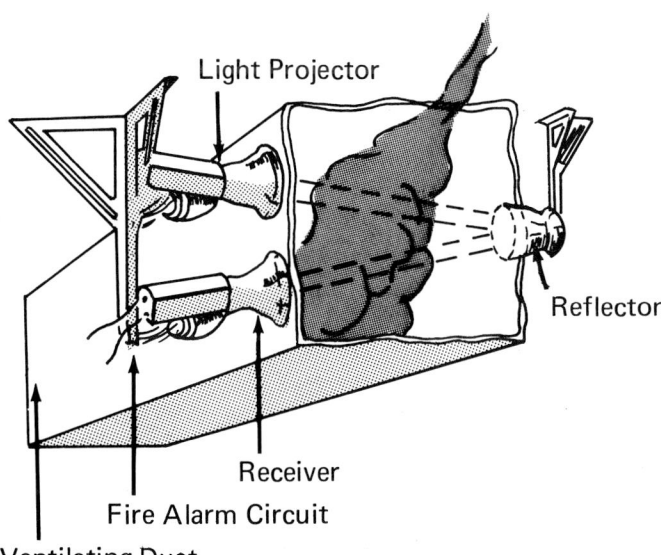

Figure 4-32 The line-type photocell can be utilized to monitor ventilation equipment. The beam can be reflected through a mirror or prism to a receiver not in line with the light source.

LINE—TYPE SMOKE DETECTOR

The refractory photocell application utilizes the photocell in an opposite application. A light beam is passed through a small chamber toward the end away from the light source. The light does not normally strike the photocell and no current is normally produced. An electronic switch working opposite of the one used for beam applications is used. The switch is open when no current is flowing.

Air is drawn through the chamber constantly. When smoke is drawn through the chamber it causes the light beam to be refracted (scattered) in random directions. A portion of the scattered light strikes the photocell, initiating a current. This current closes the electronic switch activating the alarm signal. Figure 4-33 shows the refractory-type detection principle action.

Refractory-type applications are found exclusively in spot-type detectors. This type of detector must be carefully placed to insure correct availability of air. Placement of these devices in dead air spaces will reduce their effectiveness. When installed correctly these detectors provide a reasonably reliable sensitivity detector.

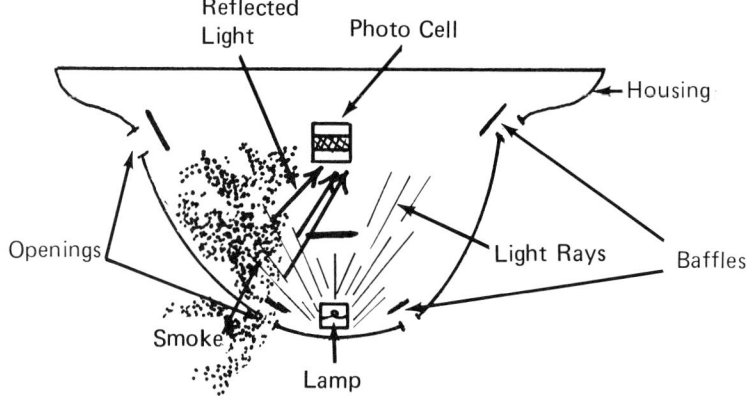

PHOTOELECTRIC SMOKE DETECTOR
SPOT—TYPE

Figure 4-33 A refractory-type photocell detector utilizes the light-scattering qualities of light instead of a direct beam to monitor for fire. This application is used exclusively in spot-type applications.

INVISIBLE-PRODUCTS-OF-COMBUSTION DETECTORS

A phenomenon of fire is the ionization of molecules as they undergo combustion. The molecules have an electron imbalance and will tend to rob electrons from other molecules. The invisible-products-of-combustion detector utilizes this phenemenon for operation.

The detector has a semi-open sensing chamber to sample the room air. A small amount of radioactive material is placed adjacent to the opening in the chamber. This radioactive material ionizes the air particles as they enter. Inside the chamber are two electrical plates. One is positively charged and one negatively charged. The negatively charged plate is closer to the radioactive source than the positive. The action of the particles ionized by the radioactive material is to free an electron, which travels to the positive plate and causes a minute current to flow between the charged plates. When ionized products of combustion enter the chamber they pick up the free electrons caused by the radioactive ionization. This causes the current flowing across the plates to cease, and initiates an alarm signal. Figure 4-34 shows the ionization principle.

Invisible-products-of-combustion detectors are commonly called ionization detectors. They share the same placement restrictions as visible-smoke detectors. Their reliability is moderate and sensitivity high.

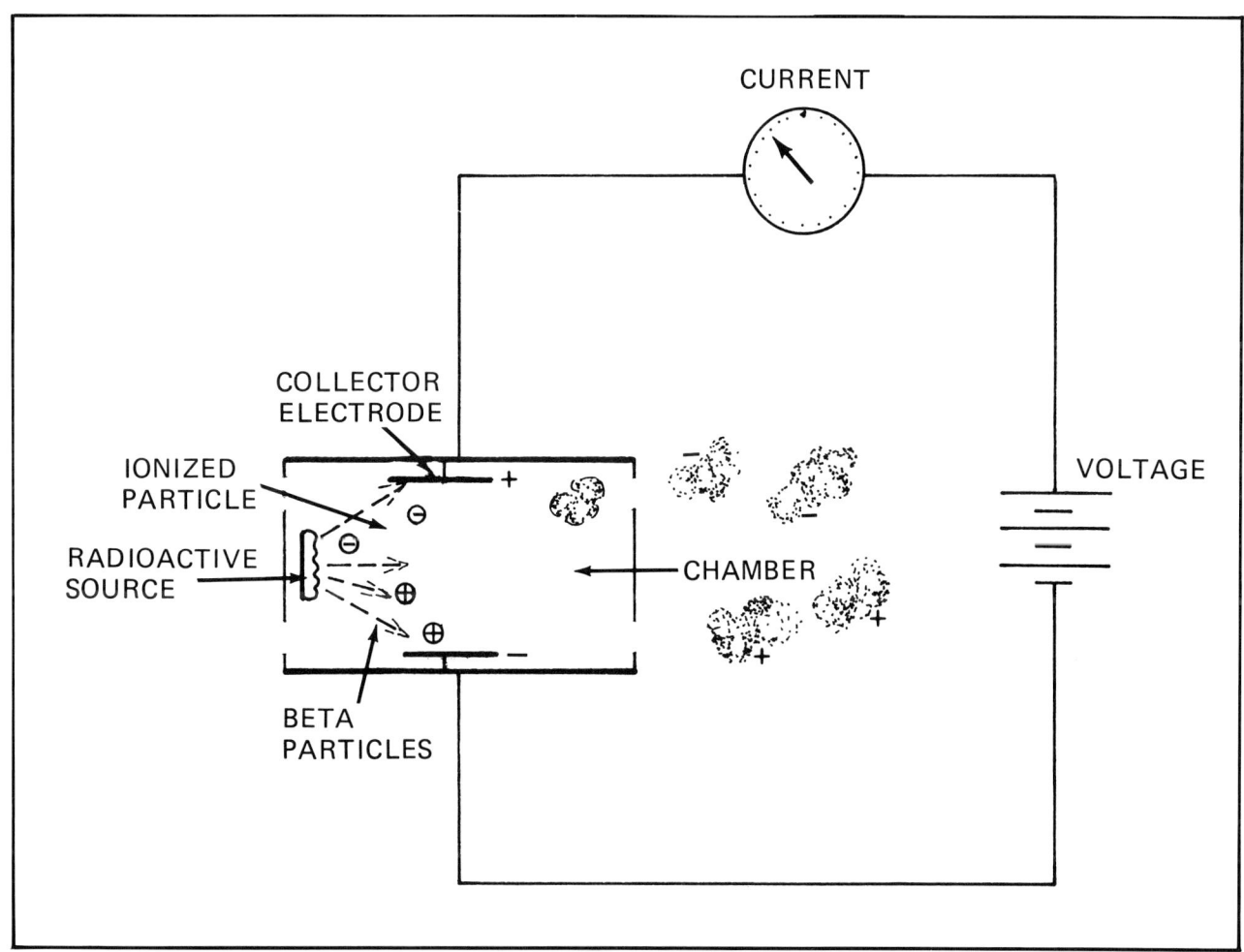

PRINCIPLE OF AN IONIZATION CHAMBER

Figure 4-34 The ionized state of smoke particles moving into the chamber neutralizes the ionizing effect of the radioactive material. This causes the current to stop flowing and sends an alarm.

LIGHT DETECTORS

Light detectors are commonly called flame detectors. There are two basic types of light detectors: The first group detects light in the ultraviolet wave spectrum; and the second group detects light in the infrared wave spectrum. These spectrums of light border on the visible range, with ultraviolet waves shorter in wave length than visible light (less than 4,000 angstroms) and infrared wave lengths somewhat longer (greater than 7,700 angstroms). Figure 4-35 shows the light spectrum.

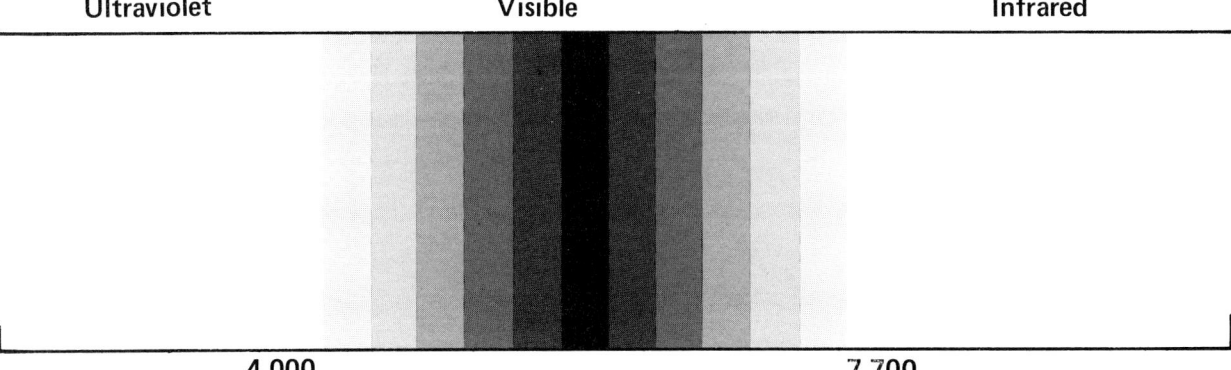

Figure 4-35 The spectrum of lightwaves that are visible range from approximately 4,000 to 7,000 angstroms.

Ultraviolet detectors electronically detect the light with wave lengths too short to be visible to the eye. These waves are generally associated with high intensity flames. A major problem with this particular detector is that sunlight is often received in the ultraviolet spectrum. This causes some stimulus-distinguishing problems for the ultraviolet detectors. In addition to sunlight, other sources such as arc welding often give off ultraviolet light.

Because of these discrimination problems of ultraviolet detectors their use is limited to systems in which the environment can be carefully controlled to exclude sources of ultraviolet light. This usually involves placement of the fire detector very close to the source of possible ignition. Although there is a need to control certain sources of light, ultraviolet detectors are widely used. Their use is finding wider acceptance as the ability to electronically distinguish or eliminate non-fire ultraviolet light is developed. Figure 4-36 pictures an ultraviolet flame detector.

Infrared detectors operate best when separated from the possible source of ignition. This makes their use in monitoring large areas very effective. It also requires the use of a dual filtering action to remove background infrared radiation. Additionally, a time delay is usually incorporated to eliminate other source of nonfire infrared radiation (Figure 4-37).

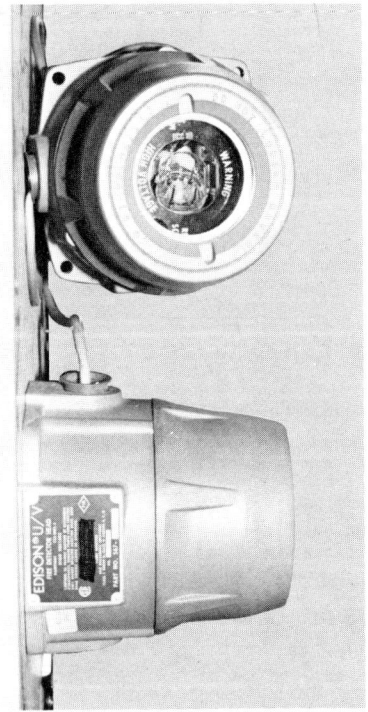

Figure 4-36 Ultraviolet detectors detect lightwaves in the spectrum below 4,000 angstroms. Generally their use is restricted to controlled environments. With the development of electronic light-discriminating circuits, their use is growing.

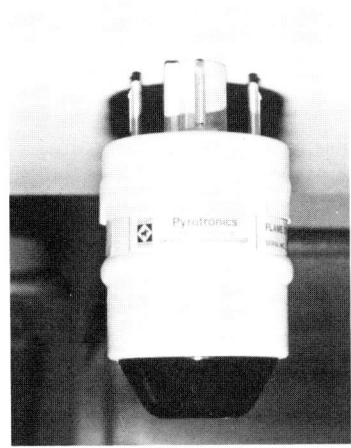

Figure 4-37 Infrared detectors are often recognized by the dark red lens on one end that reflects unwanted wave lengths.

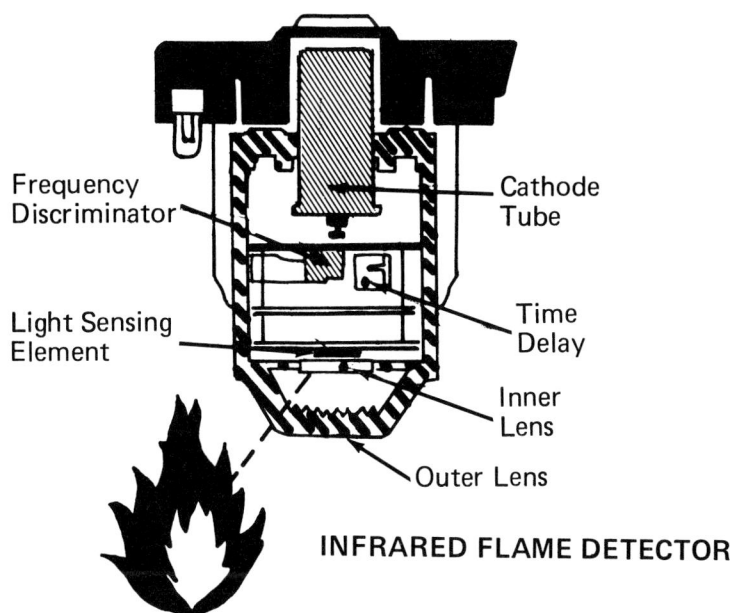

Figure 4-38 Infrared detectors detect light above 7,700 angstroms in wave length. They are a good large-area coverage detector. *(Courtesy of Pyrotronics, a subsidiary of Baker Industries)*

One method of discriminating against nonfire sources of infrared radiation is to require the additional "flickering" action of flame as a requirement before an alarm is activated. Flames typically flicker at a frequency between 50 and 30 Hz. The detector is programmed to sound an alarm only when it is receiving infrared radiation flickering in this range. Figure 4-38 is a cutaway drawing of an infrared flame detector.

Extinguishing-System Alarm-Initiating Devices

Most fixed fire extinguishing systems are equipped with some type of audible system alarm. These different devices are discussed in Sections 1 and 2 of this manual. It is important for the firefighter to understand that any fixed extinguishing system can also be interconnected to act as an alarm-initiating device for a building alarm system.

In total-flooding systems the alarm system may be set up in the presignal mode to allow occupants an opportunity to evacuate. This would also be applicable to deluge and pre-action sprinkler systems.

Telephone Dialers

A telephone dialer is a device which is connected to the telephone system from the fire alarm system of a building. The purpose of this device is to automatically dial a pre-programmed telephone number when the fire alarm system is operated. This type of system can also be used as a burglar alarm to automatically call the police department. The recorded message is replayed three or four times to assure that the dispatcher receives the entire recorded message including the correct address. At the present time, many cities and even some states have outlawed the use of automatic telephone dialers. The reason for this is that during severe thunderstorms these devices may operate at the location. Several fire departments have many additional false alarms due to this problem.

The future use of telephone dialers as a means of notifying fire and police departments is doubtful unless the technology is improved.

SUPERVISORY CIRCUITS

Alarm circuits must be supervised

For an alarm circuit to be effective, it must be reliable. This realibility necessitates the alarm circuits being able to detect faults in the system circuitry. The three basic problems in a system are open circuits caused by breaks, ground faults caused by the wire shorting out and power-source problems.

The most common method of supervising a circuit is through an end-of-the-line device. Generally, this device is a resistor or a diode. The purpose of this device is to allow a very small current to flow through the circuit at all times. This small amount of current is large enough to be monitored, but not large enough to activate the alarm relays. If the line breaks at some point, no current will flow and the monitoring device in the alarm panel will register a trouble signal.

End-of-the-line and between the circuits devices can be used for supervision

In addition to an end-of-the-line device, a monitoring device can be placed between the circuit wires. This device monitors the flow in each wire. The most well-known of these devices is the wheatstone bridge. The device acts like a small balance between the wires. The principle of the action requires that current going into the circuit from the positive side must have equal current returning on the negative side. In the event of the wire shorting out somewhere in the system, more current would leave on the positive side of the circuit than would return on the negative. This would cause the monitoring device to become unbalanced and trip, indicating a ground fault.

The power is monitored primarily on the incoming AC circuit and on the DC-power storage, or battery-circuit. There are some straight AC systems which, of course, will not have DC storage monitoring. These systems can only be used in an occupancy with emergency AC-power generation capability. The AC-power monitoring on a typical system is simply a relay placed in the AC circuit. This relay controls two separate secondary circuits. The first, when activated, switches the alarm system to the DC backup system; and the second activates the trouble-signal circuit. When AC power is flowing, the relay holds the switch of the two secondary circuits open. When the AC power fails, the relay allows the switch of the two secondary circuits to close, energizing the alternate circuits.

Monitoring of the battery is more complicated. Typically, the battery is equipped with a battery charger. The charger is controlled through a device much like an automobile's voltage regulator. This device monitors the charge, or amperage, of the battery and when the charge falls below a certain level, turns the battery charger on. When the battery is recharged, the charger is automatically turned off by the device. In addition to a charge monitoring device, a second similar device is employed. It monitors the battery charger but at a lower level. When the battery power falls below the preset level of the second device, a low-power-trouble signal is activated. The two devices working together allow the battery to be charged without sending an alarm. When the charger fails to bring the battery amperage back up and it falls below the level of the second device, a trouble signal is sent.

Most alarm panels have trouble-signal circuits that produce an audible signal in conjunction with lighting a lamp. Figure 4-39 shows a typical trouble-signal lamp arrangement. On smaller systems, only one trouble lamp is installed. The maintenance crew responding then has to determine what the source of the trouble is. On more sophisticated systems, each monitoring device will have its own annunciator lamp. Figure 4-40 shows a system with more than a "common trouble" signal. The trouble signal will be equipped with a silencer switch. This switch is to allow the repairman to turn the audible signal off while he works on the system. The audible-signal silencer switch must be wired so if the switch is in the off position and if no trouble signal is being received by the panel, the audible alarm will sound. This prevents the audible signal from inadvertently being turned or left off.

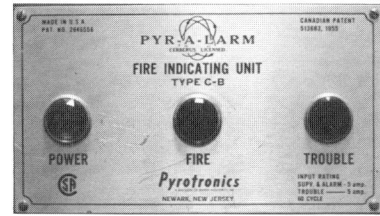

Figure 4-39 Simple detection systems will only have a few general lamps to indicate the system status. Generally the smallest number of lamps will be three. One lamp for AC power, one for trouble and one for alarm.

Figure 4-40 More complex systems will have more complicated panels. The panel may have a lamp for each monitoring circuit.

MULTIPLEX SYSTEMS (MICROPROCESSORS)

Many parts of the United States have cable television that use coaxial wire to transmit the television picture. Coaxial wire is a single wire surrounded by a flexible metal shield. This wire has the capability of handling several types of messages traveling in both directions at the same time. Because of this feature, it can be used to relay messages from a fire alarm or suppression system back to a computer or microprocessor located at the fire department or a central dispatching center. Potentially, every structure within a community having a cable could be hooked up to a central computer by means of the cable television wires. The computer or microprocessor constantly checks the alarm systems to insure that each one is in the working position. When the alarm system activates, the computer will register it, notify the proper authorities, and can print out the alarm location, time, owner and nature of the problem. It can also be programmed to print out such information as the most direct route to the scene, available man-

Total community monitoring is available with a microprocessor

power and equipment and the exact location for the alarm in relationship to the structure. This type of system is being tested in a model community in Texas. There, smoke detectors are placed in every home and are connected to a microprocessor at the local fire department.

Some modern high-rise buildings utilize a similar system where each room within the structure is connected to a central microprocessor located within the building. Upon receiving an alarm, the microprocessor can adjust the air-handling system to exhaust heat and smoke, close all fire doors, call the fire department, activate any automatic extinguishing systems and even play pre-recorded evacuation messages over the public address system. An example of just such a building is the Federal Office Building in Seattle, Washington.

Eventually, the coaxial wires will be replaced by optical fibers which can handle more signals and takes less space.

INSPECTING AND TESTING OF ALARM AND DETECTION DEVICES

Fire department personnel making inspections will find it necessary to be knowledgeable of these systems for inspection purposes. It is not within the area of responsibility of the department to maintain these systems. Industrial fire service personnel also will be required to inspect and often to test these devices. They may be required to maintain the systems, depending upon the scope of their duties.

Properly trained technicians will be needed to repair equipment

In order to maintain this equipment, personnel should be properly trained, preferably by a factory representative, on how to troubleshoot and repair the equipment. In addition, copies of the technical and parts manuals must be obtained from the manufacturer to assist in performing routine maintenance equipment failure repair.

If competent, highly trained technicians are not available to work on these devices, perhaps a service contract with an alarm company should be secured. This would insure that those working with the alarm and detection devices were familiar with their operating mechanisms.

Inspecting

A general inspection of equipment and circuit wiring should be done routinely. Inspect all circuit wiring for proper support, wear, punctures, cracks and other defects which may render the insulation ineffective. Where circuits are enclosed in conduit, inspect the conduit for solid connections and proper support. Keep all sounding devices (bells, gongs, buzzers, etc.) free from paint and dirt which may interfere with their operation.

All equipment must be kept free of dust, dirt and similar foreign materials. Recommendations for cleaning this equipment should encourage the use of a vacuum cleaner, rather than wiping the dust and having it settle on electrical contacts or relays.

Control panels, annunciator boxes, recording instruments and other devices should not have objects stored on or in them. Many units with lockable doors have ample room inside to store extra relays, rectifiers, light bulbs and test equipment. These items can foul moving parts or cause electrical shorts that can cause a failure of the system.

Keep equipment clean and uncluttered

Where batteries are used as an emergency power source, they should be checked for clean contacts. See that the float balls indicate the battery is well charged. The batteries and rectifiers should be kept free from dust and other materials.

Manual Pull Stations

During the inspection of private pull box stations, numerous items should be checked. On new installations, the inspector should check the location of pull stations to insure that they are located near main exits and all natural paths of escape. Each floor shoud have pull stations which are not more than 200 feet from any point upon the floor and each pull station must be protected from malicious tampering. On existing installations, access to the pull stations should be unobstructed, and each unit should be easy to operate. The housing should be tightly closed to prevent dust and moisture from entering the unit and disrupting service. Any damaged or cracked glass should be immediately replaced. Finally, the door to the unit should be opened to insure that all operating parts can be reached and serviced.

Detector Testing

Detectors are the most important part of the alarm system. Without them, the most elaborate system with the most sophisticated circuitry is useless. The reliability of the detector, is in fact, the reliability of the alarm system. For this reason, detectors require periodic testing. All fire detector testing shall be in accordance with NFPA Standard No. 72E, *Automatic Fire Detectors*. Three conditions indicate a need for testing: at the initial installation, after a fire and after an elapsed period of use.

Detectors must be tested to maintain system reliability

- Nonrestorable fixed-temperature detectors obviously cannot be periodically tested. Testing will destroy the detector. For this reason, tests are not required until 15 years after installation. After 15 years, two percent of the detectors must be replaced with new detectors and shipped to a nationally recognized testing laboratory for testing. Failure of any detector requires additional detectors to be submitted for testing. Cable-type line detectors must have the loop resistance tested semi-annually.

158 PRIVATE FIRE PROTECTION

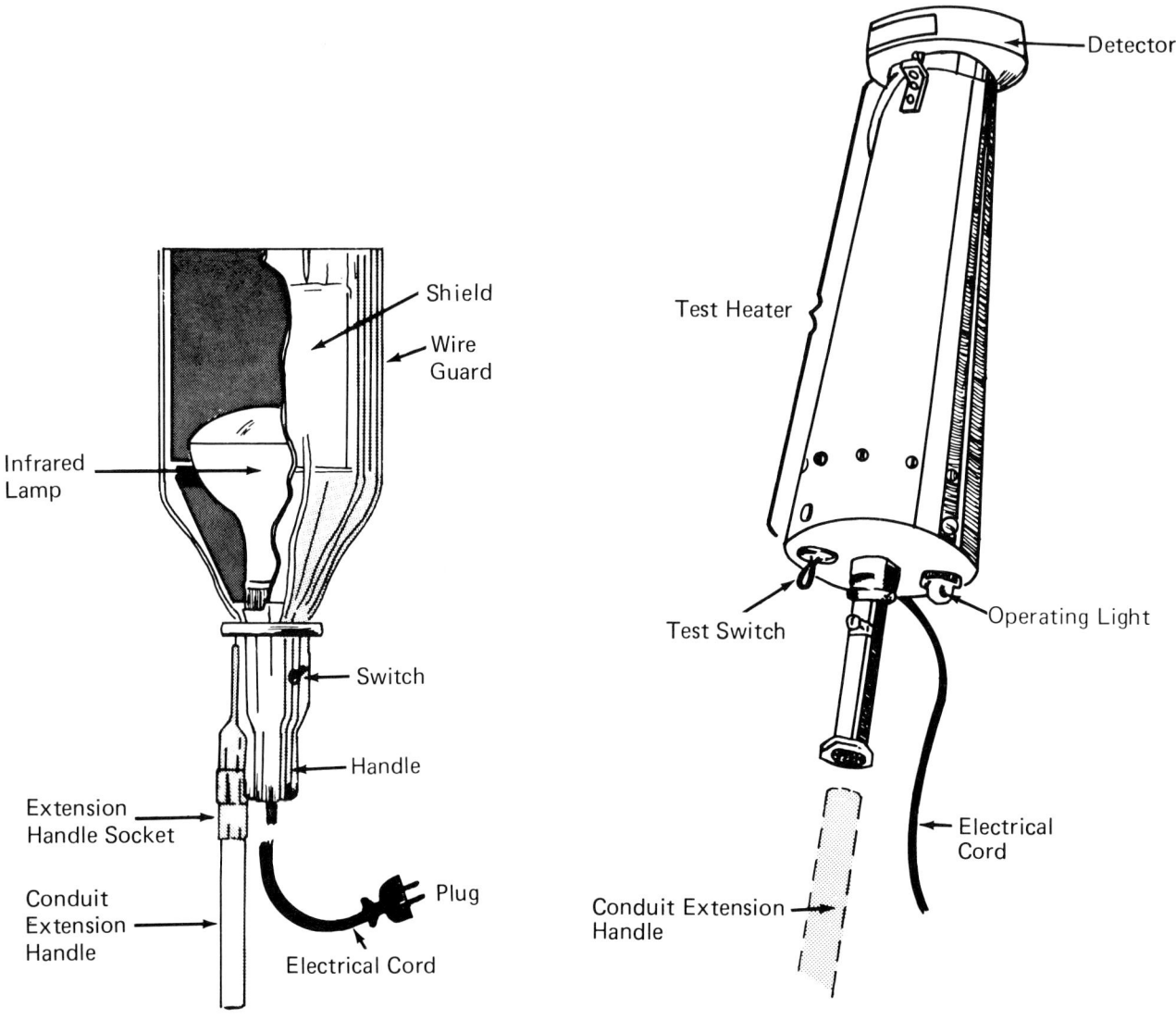

Figure 4-41 Shown are two types of detector testors. These may be purchased from the manufacturer, or may be fabricated by the local user.

- Restorable detectors all be tested with a portable heat source, such as a hairdryer or heat lamp with a temperature shield. Figure 4-41 shows two types of restorable detector test units. One detector on each signal circuit should be tested semi-annually. A different detector must be tested each time.

- Fusible link detectors with replaceable links should be tested semi-annually. This is done by removing the link and observing whether or not the contacts close. After the test, the fusible link should be replaced. It is recommended that the links be replaced with new ones at five year intervals.

- Pneumatic-type detectors may be tested with a heating device or with an approved pressure pump. The testing should be conducted on a semi-annual basis. If a pressure pump is used, the manufacturer's instructions must be followed.

- Smoke detectors should be tested semi-annually in accordance with the manufacturer's instructions. Instruments to perform performance testing and sensitivity testing are generally available from the manufacturer. Blowing cigarette smoke into the detector is not considered an acceptable test.
- Flame detectors usually are intricate and expensive pieces of equipment. Instructions from the manufacturer for test procedures and training, if obtainable, should be required by the bid specification. It is recommended that testing of the system be included in the service contract.

Detector heads must not be damaged or painted. Regardless of the type of detectors in use, the following detectors should either be replaced or sent to a recognized laboratory for testing:

Detector heads must not be painted or damaged

- Detectors on systems that are being restored to service after a period of disuse.
- Detectors that are perceptibly corroded.
- Detectors that have been painted in the field, unless they are of a type found by the testing laboratory to be unaffected by painting.
- Detectors that have been cleaned of paint.
- Detectors that have been subjected to mechanical injury or similar abuse.
- Detectors on circuits that have been subjected to surges by overvoltages or lightning damage.
- Detectors that are subjected to other conditions that may permanently affect their operation, such as grease or other deposits or corrosive atmospheres.

A permanent record of all detector tests must be maintained for a minimum period of five years. The date, detector type and location, type of test and results are the minimum information that should be included in the record.

Keep records of detector tests

Inspecting Control Equipment

The inspection of these systems includes checking of the control equipment. This consists of local annunciator panels and signal switching or transmitting devices; the printers, annunciator consoles, sounding devices and power sources at proprietary, remote and central station locations.

When necessary the inspector should witness the signal switching or transmitting device test, which will also provide an opportunity to test recording mechanisms. The initiating devices should be checked for sticking, binding or dirty contacts. The complete number of alarm signals or rounds should be sent in accordance with the design specifications and properly received.

Auxiliary devices can be checked at this time such as local evacuation alarms or air-handling equipment. All devices must be restored to proper operation after testing. Local annunicator panels can be checked during the restorable-detectors tests as previously described or by bridging contacts.

In connection with these tests, the receiving devices can be checked. The proper signal and/or number of signals should be received or recorded. Signal impulses should be definite, clear and evenly spaced to definitely identify each coded signal. There should be no sticking, binding or other irregularities. At least one complete round of printed signals should be clearly visible, unobstructed by the receiver at the end of the test. The time stamp should clearly indicate the time of a signal and should not interfere in any way with the recording device.

After testing, all devices must be returned to standby condition

All gongs for sounding alarms should be inspected to see they are clean and operable. The supervisory power source should be checked for proper operation by viewing the pilot light and/or gauge. If an emergency power supply is installed, it can be tested by interrupting the 100-volt AC power to insure that the unit will convert to battery power. After all tests are complete, the devices should be returned to their normal standby condition.

APPENDIX A
The Metric System Of Measurement

Since the development of the metric system in the eighteenth century, the vast majority of the world's countries have gone metric. In fact, only six countries, in 1974, were uncommitted to the metric system.

There are two basic reasons for the conversion of countries to the metric system. The first reason for going metric is for simplicity and efficiency. The metric system is based upon simple powers of ten where mass, volume and length are all interrelated in the system of measures. The fact that the power of ten is used in metrics makes computations simple, with a reduction in errors and a savings of time.

The second fundamental reason for adopting the metric system of measurement is fairly obvious considering the fact that almost all countries are currently using metrification The simple economics of world trade displays the advantages and necessity of using a common system of measurement. Since the majority of countries are using metrics, they supply products in metric measurements. These same countries, as consumers, will often demand products in metrics. In order to remain economically competitive in the world market the need for metrification is evident.

Since its inception, the metric system of measurement has not been a static one. The system has continued to be refined and, as a result, the ultimate in metrification has been developed. The International System of Units, simply referred to as SI, was devised in 1960 by the metric nations. Already adopted by more than 30 countries, this system embodies the latest in metric study.

UNITS OF METRIC SYSTEM

Length

The basic unit of the metric system is the meter. In establishing the length of the meter, the developers wanted to use a basic fact of nature. After several changes the meter was finally defined as "1,650,763.73 wave lengths of the orange-red line of krypton 86". This definition is based on a physical condition which will always be the same and is a true standard of length. From the basic unt of the meter, the system of lengths is simply expanded in multiples of tens in either direction. By using powers of ten, confusing fractions are eliminated since all lengths can be simply expressed using decimals. The length measurements in metrics and English equivalents are:

MULTIPLIER		UNIT OF LENGTH	SYMBOL		OUR EQUIVALENT
1/1000 meter	=	1 millimeter	mm	=	000.0393701 inch
10 millimeters	=	1 centimeter	cm	=	000.3937000 inch
10 centimeters	=	1 decimeter	dm	=	003.9370000 inches
10 decimeters	=	1 meter	m	=	001.0936100 yards
10 meters	=	1 decameter	dam	=	010.9360000 yards
10 decameters	=	1 hectometer	hm	=	109.3000000 yards
10 hectometers	=	1 kilometer	km	=	000.6213710 miles

LENGTH

TO CONVERT	TO	MULTIPLY BY
inches	centimeters (cm)	2.5400
feet	meters (m)	0.3048
yards	meters (m)	0.9144
miles	kilometers (km)	1.8520

As you will notice, the word meter is used in developing all other words. The prefixes used are either Greek or Latin words or roots. The same prefixes are used in developing other measurements — again, a simplification. Since metrics is in powers of ten, it is extremely simple to make other conversions. For instance, how many millimeters are equal to an inch? Knowing that 2.54 cm = 1 inch and knowing that 10 mm = 1 cm, it is simple to deduce 1 inch = 25.4 mm.

Metric Squares

The method for devising square measures is simple. For example, begin with a square land plot measuring one meter per side. This is a square meter (m^2). A square this size is also equal to one which measures 10 dm by 10 dm (since 10 dm = 1 m) and therefore is equal to 10 dm × 10 dm = 100 square decimeters (dm^2). We can then say that $1\ m^2 = 100\ dm^2$.

MULTIPLIER		UNIT OF AREA	SYMBOL		OUR EQUIVALENT
—	=	1 sq. millimeter	mm^2	=	00.00155 sq. in.
100 mm^2	=	1 sq. centimeter	cm^2	=	00.15500 sq. in.
100 cm^2	=	1 sq. decimeter	dm^2	=	15.50000 sq. in.
100 dm^2	=	1 sq. meter	m^2	=	01.19600 sq. yd.
100 m^2	=	1 sq. decameter	dam^2	=	00.02500 sq. yd.
100 dam^2	=	1 sq. hectometer	hm^2	=	02.47100 sq. yd.
100 hm^2	=	1 sq. kilometer	km^2	=	00.38600 sq. yd.

One hm^2 is often referred to as a hectare (ha).

AREA

TO CONVERT	TO	MULTIPLY BY
square inches	square centimeters (cm^2)	6.452
square feet	square centimeters (cm^2)	929.030
square yards	square meters (m^2)	0.836
acres	hectares	0.405
square miles	hectares	258.999
square miles	square kilometers (km^2)	2.590

Appendix A **163**

Metric Cubes

To measure volume or capacity, develop a cube 1 meter by 1 meter by 1 meter. Its volume is a cubic meter (m^3). This cube is the same size as a cube measuring 10 dm to a side, and therefore has the volume of 10 dm $\times$ 10 dm $\times$ 10 dm = 100 dm^3. Consequently, there are 1000 dm^3 in 1 m^3. A cubic metric unit is always 1000 times larger than the next smallest one.

MULTIPLIER		UNIT OF VOLUME	SYMBOL		OUR EQUIVALENT
	=	1 cubic millimeter	mm^3	=	6.1×10^5 cubic inch
1000 mm^3	=	1 cubic centimeter	cm^3	=	00.06100 cubic inch
1000 cm^3	=	1 cubic decimeter	dm^3	=	61.02300 cubic inches
1000 dm^3	=	1 cubic meter	m^3	=	01.30795 cubic yards
1000 m^3	=	1 cubic decameter	dam^3	=	
1000 dam^3	=	1 cubic hectometer	hm^3	=	
1000 hm^3	=	1 cubic kilometer	km^3	=	

VOLUME: CUBIC

TO CONVERT	TO	MULTIPLY BY
cubic inches	cubic centimeters (cm^3)	16.3870
cubic feet	cubic decimeters (dm^3)	28.3170
cubic yards	cubic meters (m^3)	00.7645

VOLUME AND CAPACITY

The basic unit of measure for volume or capacity is the liter. The liter is defined as 1 kilogram of pure water at 4°C. The liter is also equal in volume to one cubic decimeter (1 dm^3); hence the interrelationship of measurement of volume and length. To expand the metric volume, all that is needed is to expand the liter by multiples of 10 in either direction and apply the normal prefixes.

MULTIPLIER		UNIT OF CAPACITY	SYMBOL		EQUIVALENT LINEAR MEASURES
1/1000 liter	=	1 milliliter	ml	=	1 cubic centimeter
10 milliliters	=	1 centiliter	cl	=	10 cubic centimeters
10 centiliters	=	1 deciliter	dl	=	100 cubic centimeters
10 deciliters	=	1 liter	l	=	1 cubic decimeter
10 liters	=	1 decaliter	dal	=	10 cubic decimeters
10 decaliters	=	1 hectoliter	hl	=	100 cubic decimeters
10 hectoliters	=	1 kiloliter	kl	=	1 cubic meter

The English conversions are as follows:

TO CONVERT	TO	MULTIPLY BY
ounces	milliliters (ml)	296.0000
pints	liters (l)	000.4732
quarts	liters (l)	000.9464
gallons	liters (l)	003.7850

The most commonly used conversion will be in converting gallons per minute to liters per minute: 1 gpm = 3.785 liters/minute (l/min.)

METRIC MASS

The basic metric unit for mass measurement is the gram. It has already been discussed that 1 cubic centimeter = 1 milliliter and furthermore 1 milliliter has a volume equal to that of 1 gram of water. Again the measurements are interrelated. By multiples of ten, the metrics for mass include:

MULTIPLIER		UNIT OF MASS	SYMBOL		OUR EQUIVALENT
1/1000 gram	=	1 milligram	mg	=	0.015 grain
10 milligrams	=	1 centigram	cg	=	0.154 grains
10 centigrams	=	1 decigram	dg	=	1.543 grains
10 decigrams	=	1 gram	g	=	0.035 ounces
10 grams	=	1 decagram	dag	=	0.353 ounces
10 decagrams	=	1 hectogram	hg	=	3.527 ounces
10 hectograms	=	1 kilogram	kg	=	2.205 pounds

One kilogram is further defined as the mass of 1 liter of pure water at 4°Celsius.

1 kg = 1000 g   but 1 liter of water has a mass of 1 kg and 1000 ml = 1 dm^3
1 l = 1000 ml   therefore, 1 dm^3 of water has a mass of 1000 g.

TO CONVERT	TO	MULTIPLY BY
ounces	grams (g)	028.350
pounds	grams (g)	453.592
pounds	kilograms (kg)	00.4540

MEASURING TEMPERATURE

The Celsius, or Centigrade, system of heat measurement has also been adopted by most countries using metrics. In this system, the freezing point of water is 0° and the boiling point of water is 100°. The difference is 100° which fits in well with the decimal notation of the metric system. To convert Celsius to Fahrenheit, use °C = 5/9 (°F - 32). To convert the other way, use °F = 9/5 (°C) + 32.

PASCAL

One other metric unit that is particularly important to firefighters is the unit to measure pressure, the pascal (Pa).

$$1 \text{ psi} = 6894.757 \text{ Pa, so } 1 \text{ psi} = 6.895 \text{ kPa}$$

Metric Equivalents

PAGE	ENGLISH		METRIC
	INTRODUCTION		
2	150 to 2,000 pounds	=	68.1 to 908 kg
	SECTION 1		
11	1/2 inch	=	1.27 cm
11	3/8 inch	=	.95 cm
11	17/32 inch	=	1.35 cm
11	130 to 168 sq. feet	=	12.08 to 15.6 m^2
11	200 sq. feet	=	18.58 m^2
11	65 to 100 sq. feet	=	6.04 to 9.29 m^2
11	130 sq. feet	=	12.08 m^2
11	15 feet	=	4.6 meters
11	100 sq. feet	=	9.29 m^2
11	90 sq. feet	=	8.4 m^2
11	12 feet	=	3.7 meters
12	1/2 inch	=	1.27 cm
15	135 to 170°F	=	57.2 to 76.7°C
15	175 to 225°F	=	79.4 to 107.2°C
15	250 to 300°F	=	121 to 148.9°C
15	325 to 375°F	=	162.8 to 190.6°C
15	400 to 475°F	=	204.4 to 246.1°C
15	500 to 575°F	=	260 to 301.7°C

PAGE	ENGLISH		METRIC
21	15 psi	=	103.43 kPa
21	35 feet	=	10.67 m
21	75 feet	=	22.86 m
24	150 psi	=	1034.3 kPa
25	1 1/2 inch	=	3.81 cm
25	2 1/2 inch	=	6.35 cm
25	150 psi	=	1034.3 kPa
25	175 to 200 psi	=	1207 to 1379 kPa
34	18 inch	=	.45 m
36	2 inch	=	5.08 cm
37	2 inch	=	5.08 cm
39	15 psi	=	103.43 kPa
39	50 psi	=	344.75 kPa
40	500 gallons	=	1892.5 l
40	1 psi	=	6.9 kPa
40	2 psi	=	13.8 kPa
42	40°F	=	4.4°C
42	15 to 20 psi	=	103.43 to 206.86 kPa
44	2 inch	=	5.08 cm
45	2 inch	=	5.08 cm
47	2 inch	=	5.08 cm
48	2 inch	=	5.08 cm
52	2 inch	=	5.08 cm
54	2 inch	=	5.08 cm

SECTION 2

PAGE	ENGLISH		METRIC
60	700°F	=	371°C
60	600°F	=	315°C
64	300 psi	=	2068.2 kPa
64	0°F	=	−17.8°C
64	500 pounds to 125 tons	=	227 kg to 137.5 tonnes
64	325 psi	=	2240.88 kPa
64	50 to 100 pounds	=	22.7 kg to 45.4 kg
64	850 psi	=	5860.75 kPa
64	70°F	=	21°C
65	0°F	=	−17.8°C
65	300 psi	=	2068.2 kPa
65	850 psi	=	5860.75 kPa
65	70°F	=	21°C
65	100 pounds	=	45.4 kg
70	5 to 250 pounds	=	2.27 to 113.5 kg
82	100°F	=	42°C
82	20°F	=	−7°C
82	120°F	=	49°C

PAGE	ENGLISH		METRIC
82	−20°F	=	−29°C
83	35°F	=	2°C
83	120°F	=	48.9°C
85	6 feet	=	1.8 m
90	60 to 180 gpm	=	227 to 681 l/min
90	1,000 gpm	=	3,785 l/min
95	250 to 4,000 gpm	=	946 to 15,140 l/min
101	2 feet	=	.61 m
101	24 feet	=	7.32 m
102	212°F	=	100°C

SECTION 3

PAGE	ENGLISH		METRIC
105	2 1/2 inch	=	6.35 cm
106	2 1/2 inch	=	6.35 cm
107	500 gpm	=	1892.5 l/min
107	250 gpm	=	946.25 l/min
107	65 psi	=	448.18 kPa
107	100 gpm	=	378.5 l/min
107	100 psi	=	689.5 kPa
107	150 psi	=	1034 kPa
108	275 feet	=	83.8 m
108	325 feet	=	99.06 m
108	200 feet	=	60.96 m
108	125 feet	=	38.1 m
108	550 feet	=	167.64 m
109	550 feet	=	167.64 m
110	6 feet	=	1.8 m
110	200 psi	=	1379 kPa
110	75 psi	=	517.13 kPa
111	25 psi	=	172.38 kPa
112	25 psi	=	172.38 kPa
112	150 feet	=	45.72 m
112	2½ × 1½ × 1½ inch	=	6.35 × 3.81 × 3.81 cm
112	3 inch	=	7.62 cm
112	1 1/2 inch	=	3.81 cm
112	2 1/2 inch	=	6.35 cm
113	3 1/2 inch	=	6.35
113	1 1/2 inch	=	3.81 cm
113	3 inch	=	7.62 cm
113	2½ × 2½ × 2½ inch	=	6.35 × 6.35 × 6.35 cm
113	100 psi	=	689.5 kPa
114	2 1/2 inch	=	6.35 cm
117	2 1/2 gallon	=	9.46 l
117	1 1/4 gallon	=	4.73 l

PAGE	ENGLISH		METRIC
117	2 inch	=	5.08 cm
117	18 3/4 inch	=	47.63 cm
117	24 inch	=	60.96 cm
	28 3/4 inch	=	73.03 cm
117	31 1/4 inch	=	79.38 cm
117	36 inch	=	91.44 cm
117	44 1/2 inch	=	113.03 cm
117	60 inch	=	152.4 cm
117	72 inch	=	182.88 cm
117	84 inch	=	213.13 cm
117	8 feet	=	2.44 m
117	10 feet	=	3.05 m
117	12 feet	=	3.66 m
117	14 feet	=	4.27 m
117	17 feet	=	5.19 m
117	1 gallon	=	3.79 l
117	2 gallons	=	7.58 l
117	3 gallons	=	11.37 l
117	4 gallons	=	15.16 l
117	6 gallons	=	22.74 l
117	10 pounds	=	4.536 kg
117	20 pounds	=	9.072 kg
117	40 pounds	=	18.144 kg
117	60 pounds	=	27.216 kg
117	6 pounds	=	2.722 kg
117	12 pounds	=	5.443 kg
117	18 pounds	=	8.165 kg
117	24 pounds	=	10.886 kg
117	36 pounds	=	16.33 kg
117	2' 10" × 5' 8"	=	86.36 × 172.72 cm
117	4" × 8"	=	121.92 × 243.84 cm
117	4' 11" × 9' 9 1/2"	=	1149.86 × 298.45 cm
117	6' × 10' 8"	=	182.88 × 325.12 cm
117	6' 11" × 13' 11"	=	210.82 × 424.18 cm
118	10 sq. feet	=	.93 m^2
118	8 inch	=	20.32 cm
118	25 sq. feet	=	2.32 m^2
118	2 1/2 sq. feet	=	2322.58 cm^2
118	5 sq. feet	=	4645.15 cm^2
118	12 1/2 sq. feet	=	1.16 m^2
118	50 sq. feet	=	4.65 m^2
118	75 sq. feet	=	6.97 m^2
118	100 sq. feet	=	9.29 m^2
118	150 sq. feet	=	13.94 m^2

PAGE	ENGLISH		METRIC
118	200 sq. feet	=	18.58 m^2
118	300 sq. feet	=	27.87 m^2
118	400 sq. feet	=	37.16 m^2
118	600 sq. feet	=	55.74 m^2
118	800 sq. feet	=	74.32 m^2
118	1,200 sq. feet	=	111.48 m^2
118	1,600 sq. feet	=	148.64 m^2
118	3 1/4 gallons	=	12.32 l
118	6 1/4 gallons	=	23.69 l
118	15 1/2 gallons	=	58.75 l
118	65 gallons	=	246.35 l
118	95 gallons	=	360.05 l
118	125 gallons	=	473.75 l
118	190 gallons	=	720.1 l
118	250 gallons	=	947.5 l
118	375 gallons	=	1421.25 l
118	500 gallons	=	1895 l
118	750 gallons	=	2842.5 l
118	1,000 gallons	=	3790 l
118	1,500 gallons	=	5685 l
118	2,000 gallons	=	7580 l
120	40 pounds	=	18.15 kg
120	3 1/2 feet	=	107 m
120	4 inch	=	10.16 cm
120	3,000 sq. feet	=	27.88 m
120	3,000 sq. feet	=	278.71 m^2
120	4,000 sq. feet	=	371.61 m^2
120	4,500 sq. feet	=	418.06 m^2
120	6,000 sq. feet	=	557.41 m^2
120	9,000 sq. feet	=	836.12 m^2
120	11,250 sq. feet	=	1045.15 m^2
121	50 feet	=	15.25 m^2
121	30 feet	=	9.15 m

SECTION 4

142	32°F	=	0°C
143	100°F	=	37.8°C
143	135°F	=	57.2°C
143	150°F	=	65.6°C
143	160°F	=	71°C
143	175°F	=	79.4°C
143	225°F	=	107.2°C

PAGE	ENGLISH		METRIC
143	250°F	=	121°C
143	300°F	=	148.9°C
143	325°F	=	162.8°C
143	360°F	=	182.2°C
146	5°F	=	−15°C
146	135°F	=	57.2°C
146	140°F	=	60°C
157	200 feet	=	61 m

IFSTA MANUALS AND FPP PRODUCTS

For a current catalog describing these and other products call or write your local IFSTA distributor or Fire Protection Publications, IFSTA Headquarters, Oklahoma State University, Stillwater, OK 74078-0118.
Phone: 1-800-654-4055

FIRE DEPARTMENT AERIAL APPARATUS
includes information on the driver/operator's qualifications; vehicle operation; types of aerial apparatus; positioning, stabilizing, and operating aerial devices; tactics for aerial devices; and maintaining, testing, and purchasing aerial apparatus. Detailed appendices describe specific manufacturers' aerial devices. 1st Edition (1991), 416 pages, addresses NFPA 1002.

STUDY GUIDE FOR AERIAL APPARATUS
The companion study guide in question and answer format. 1st Edition (1991), 152 pages.

AIRCRAFT RESCUE AND FIRE FIGHTING
comprehensively covers commercial, military, and general aviation. All the information you need is in one place. Subjects covered include: personal protective equipment, apparatus and equipment, extinguishing agents, engines and systems, fire fighting procedures, hazardous materials, and fire prevention. Over 240 photographs and two-color illustrations. It also contains a glossary and review questions with answers. 3rd Edition (1992), 272 pages, addresses NFPA 1003.

BUILDING CONSTRUCTION RELATED TO THE FIRE SERVICE
helps firefighters become aware of the many construction designs and features of buildings found in a typical first alarm district and how these designs serve or hinder the suppression effort. Subjects include construction principles, assemblies and their resistance to fire, building services, door and window assemblies, and special types of structures. 1st Edition (1986), 166 pages, addresses NFPA 1001 and NFPA 1031, levels I & II.

CHIEF OFFICER
lists, explains, and illustrates the skills necessary to plan and maintain an efficient and cost-effective fire department. The combination of an ever-increasing fire problem, spiraling personnel and equipment costs, and the development of new technologies and methods for decision making requires far more than expertise in fire suppression. Today's chief officer must possess the ability to plan and administrate as well as have political expertise. 1st Edition (1985), 211 pages, addresses NFPA 1021, level VI.

SELF-INSTRUCTION FOR CHIEF OFFICER
The companion study guide in question and answer format. 1st Edition, 142 pages.

FIRE DEPARTMENT COMPANY OFFICER
focuses on the basic principles of fire department organization, working relationships, and personnel management. For the firefighter aspiring to become a company officer, or a company officer wishing to improve management skills, this manual helps develop and improve the necessary traits to effectively manage the fire company. 2nd Edition (1990), 278 pages, addresses NFPA 1021, levels I, II, & III.

COMPANY OFFICER STUDY GUIDE
The companion study guide in question and answer format. Includes problem applications and case studies. 1st Edition (1991), 256 pages.

ESSENTIALS OF FIRE FIGHTING
is the "bible" on basic firefighter skills and is used throughout the world. The easy-to-read format is enhanced by 1,500 photographs and illustrations. Step-by-step instructions are provided for many fire fighting tasks. Topics covered include: personal protective equipment, building construction, firefighter safety, fire behavior, portable extinguishers, SCBA, ropes and knots, rescue, forcible entry, ventilation, communications, water supplies, fire streams, hose, fire cause determination, public fire education and prevention, fire suppression techniques, ladders, salvage and overhaul, and automatic sprinkler systems. 3rd Edition (1992), addresses NFPA 1001.

STUDY GUIDE FOR 3rd EDITION OF ESSENTIALS OF FIRE FIGHTING
The companion learning tool for the new 3rd edition of the manual. It contains questions and answers to help you learn the important information in the book. 1st Edition (1992).

PRINCIPLES OF EXTRICATION
leads you step-by-step through the procedures for disentangling victims from cars, buses, trains, farm equipment, and industrial situations. Fully illustrated with color diagrams and more than 500 photographs. It includes rescue company organization, protective clothing, and evaluating resources. Review questions with answers at the end of each chapter. 1st Edition (1990), 400 pages.

FIRE CAUSE DETERMINATION
gives you the information necessary to make on-scene fire cause determinations. You will know when to call for a trained investigator, and you will be able to help the investigator. It includes a profile of firesetters, finding origin and cause, documenting evidence, interviewing witnesses, and courtroom demeanor. 1st Edition (1982), 159 pages, addresses NFPA 1021, Fire Officer I, and NFPA 1031, levels I & II.

FIRE SERVICE FIRST RESPONDER
provides the information needed to evaluate and treat patients with serious injuries or illnesses. It familiarizes the reader with a wide variety of medical equipment and supplies. **First Responder** applies to safety, security, fire brigade, and law enforcement personnel, as well as fire service personnel, who are required to administer emergency medical care. 1st Edition (1987), 340 pages, addresses NFPA 1001, levels I & II, and DOT First Responder.

FORCIBLE ENTRY
reflects the growing concern for the reduction of property damage as well as firefighter safety. This comprehensive manual contains technical information about forcible entry tactics, tools, and methods, as well as door, window, and wall construction. Tactics discuss the degree of danger to the structure and leaving the building secure after entry. Includes a section on locks and through-the-lock entry. Review questions and answers at the end of each chapter. 7th Edition (1987), 270 pages, helpful for NFPA 1001.

GROUND COVER FIRE FIGHTING PRACTICES

explains the dramatic difference between structural fire fighting and wildland fire fighting. Ground cover fires include fires in weeds, grass, field crops, and brush. It discusses the apparatus, equipment, and extinguishing agents used to combat wildland fires. Outdoor fire behavior and how fuels, weather, and topography affect fire spread are explained. The text also covers personnel safety, management, and suppression methods. It contains a glossary, sample fire operation plan, fire control organization system, fire origin and cause determination, and water expansion pump systems. 2nd Edition (1982), 152 pages.

FIRE SERVICE GROUND LADDER PRACTICES

is a "how to" manual for learning how to handle, raise, and climb ground ladders; it also details maintenance and service testing. Basic information is presented with a variety of methods that allow the readers to select the best method for their locale. The chapter on Special Uses includes: ladders as a stretcher, a slide, a float drag, a water chute, and more. The manual contains a glossary, review questions and answers, and a sample testing and repair form. 8th Edition (1984), 388 pages, addresses NFPA 1001.

HAZARDOUS MATERIALS FOR FIRST RESPONDERS

provides basic information on hazardous materials with sections on site management and decontamination. It includes a description of various types of materials, their characteristics, and containers. The manual covers the effects of weather, topography, and environment on the behavior of hazardous materials and control efforts. Pre-incident planning and post-incident analysis are covered. 1st Edition (1988), 357 pages, addresses NFPA 472, 29 CFR 1910.120 and NFPA 1001.

STUDY GUIDE FOR HAZARDOUS MATERIALS FOR FIRST RESPONDERS

The companion study guide in question and answer format also includes case studies that simulate incidents. 1st Edition (1989), 208 pages.

HAZARDOUS MATERIALS: MANAGING THE INCIDENT

takes you beyond the basic information found in **Hazardous Materials for First Responders**. Directed to the leader/commander, this manual sets forth basic practices clearly and comprehensively. Charts, tables, and checklists guide you through the organization and planning stages to decontamination. This text, along with the accompanying workbook and instructor's guide, provides a comprehensive learning package. 1st Edition (1988), 206 pages, helpful for NFPA 1021.

STUDENT WORKBOOK FOR HAZARDOUS MATERIALS: MANAGING THE INCIDENT

provides questions and answers to enhance the student's comprehension and retention. 1st Edition (1988), 176 pages.

INSTRUCTOR'S GUIDE FOR HAZARDOUS MATERIALS: MANAGING THE INCIDENT

provides lessons based on each chapter, adult learning tips, and appendices of references and suggested audio visuals. 1st Edition (1988), 142 pages.

HAZ MAT RESPONSE TEAM LEAK AND SPILL GUIDE

contains articles by Michael Hildebrand reprinted from *Speaking of Fire*'s popular Hazardous Materials Nuts and Bolts series. Two additional articles from *Speaking of Fire* and the hazardous material incident SOP from the Chicago Fire Department are also included. 1st Edition (1984), 57 pages.

EMERGENCY OPERATIONS IN HIGH-RACK STORAGE

is a concise summary of emergency operations in the high-rack storage area of a warehouse. It explains how to develop a pre-emergency plan, what equipment will be necessary to implement the plan, type and amount of training personnel will need to handle an emergency, and interfacing with various agencies. Includes consideration questions, points not to be overlooked, and trial scenarios. 1st Edition (1981), 97 pages.

HOSE PRACTICES

reflects the latest information on modern fire hose and couplings. It is the most comprehensive single source about hose and its use. The manual details basic methods of handling hose, including large diameter hose. It is fully illustrated with photographs showing loads, evolutions, and techniques. This complete and practical book explains the national standards for hose and couplings. 7th Edition (1988), 245 pages, addresses NFPA 1001.

FIRE PROTECTION HYDRAULICS AND WATER SUPPLY ANALYSIS

covers the quantity and pressure of water needed to provide adequate fire protection, the ability of existing water supply systems to provide fire protection, the adequacy of a water supply for a sprinkler system, and alternatives for deficient water supply systems. 1st Edition (1990), 340 pages.

INCIDENT COMMAND SYSTEM (ICS)

was developed by a multiagency task force. Using this system, fire, police, and other government groups can operate together effectively under a single command. The system is modular and can be used to meet the requirements of both day-to-day and large-incident operations. It is the approved basic command system taught at the National Fire Academy. 1st Edition (1983), 220 pages, helpful for NFPA 1021.

INDUSTRIAL FIRE PROTECTION

is designed for the person charged with the responsibility of developing, implementing, and coordinating fire protection. A "must read" for fire service personnel who will coordinate with industry/business for pre-incident planning. The text includes guidelines for establishing a company policy, organization and planning for the emergency, establishing a fire prevention plan, incipient fire fighting tactics, an overview of interior structural fire fighting, and fixed fire fighting systems. 1st Edition (1982), 207 pages, written for 29 CFR. 1910, Subpart L, and helpful for NFPA 1021 and NFPA 1031.

FIRE INSPECTION AND CODE ENFORCEMENT

provides a comprehensive, state-of-the-art reference and training manual for both uniformed and civilian inspectors. It is a comprehensive guide to the principles and techniques of inspection. Text includes information on how fire travels, electrical hazards, and fire resistance requirements. It covers storage, handling, and use of hazardous materials; fire protection systems; and building construction for fire and life safety. 5th Edition (1987), 316 pages, addresses NFPA 1001 and NFPA 1031, levels I & II.

STUDY GUIDE FOR FIRE INSPECTION AND CODE ENFORCEMENT

The companion study guide in question and answer format with case studies. 1st Edition (1989), 272 pages.

FIRE SERVICE INSTRUCTOR

explains the characteristics of a good instructor, shows you how to determine training requirements, and teach to the level of your class. It discusses the types, principles, and procedures of teaching and learning, and covers the use of effective training aids and devices. The purpose and principles of testing as well as test construction are covered. Included are chapters on safety, legal considerations, and computers. 5th Edition (1990), 325 pages, addresses NFPA 1041, levels I & II.

LEADERSHIP IN THE FIRE SERVICE
was created from the series of lectures given by Robert F. Hamm to assist in leadership development. It provides the foundation for getting along with others, explains how to gain the confidence of your personnel, and covers what is expected of an officer. Included is information on supervision, evaluations, delegating, and teaching. Some of the topics include: the successful leader today, a look into the past may reveal the future, and self-analysis for officers. 1st Edition (1967), 132 pages.

FIRE SERVICE ORIENTATION AND INDOCTRINATION
relates the traditions, history, and organization of the fire service. It includes operation of the fire department, responsibilities and duties of firefighters, and the function of fire department companies. This exciting and informative text is for anyone dealing with the fire service who needs a basic understanding and overview. The perfect book for new or prospective members, buffs, your congressman or council members, fire service sales personnel, and industrial brigades. 2nd Edition (1984), 187 pages, addresses NFPA 1001.

PRIVATE FIRE PROTECTION AND DETECTION
introduces the firefighter, inspection personnel, brigade/safety member, insurance inspector/investigator, or fire protection student to fixed systems, extinguishers, and detection. It covers the various types of equipment, their installation, maintenance, and testing. Systems discussed are: wet-pipe, dry-pipe, pre-action, deluge, residential, carbon dioxide, Halogen, dry- and wet-chemical, foam, standpipe, and fire extinguishers. 1st Edition (1979), 170 pages, addresses NFPA 1001 and NFPA 1031, levels I & II.

PUBLIC FIRE EDUCATION
provides valuable information for ending public apathy and ignorance about fire. This manual gives you the knowledge to plan and implement fire prevention campaigns. It shows you how to tailor the individual programs to your audience as well as the time of year or specific problems. It includes working with the media, resource exchange, and smoke detectors. 1st Edition (1979), 169 pages, helpful for NFPA 1021 and 1031.

FIRE DEPARTMENT PUMPING APPARATUS
is the Driver/Operator's encyclopedia on operating fire pumps and pumping apparatus. It covers pumpers, tankers (tenders), brush apparatus, and aerials with pumps. This comprehensive volume explains safe driving techniques, getting maximum efficiency from the pump, and basic water supply. It includes specification writing, apparatus testing, and extensive appendices of pump manufacturers. 7th Edition (1989), 374 pages, addresses NFPA 1002.

STUDY GUIDE FOR PUMPING APPARATUS
The companion study guide in question and answer format. 1st Edition (1990), 100 pages.

FIRE SERVICE RESCUE PRACTICES
is a comprehensive training text for firefighters and fire brigade members that expands proficiency in moving and removing victims from hazardous situations. This extensively illustrated manual includes rescuer safety, effects of rescue work on victims, rescue from hazardous atmospheres, trenching, and outdoor searches. 5th Edition (1981), 262 pages, addresses NFPA 1001.

RESIDENTIAL SPRINKLERS A PRIMER
outlines United States residential fire experience, system components, engineering requirements, and issues concerning automatic and fixed residential sprinkler systems. Written by Gary Courtney and Scott Kerwood and reprinted from *Speaking of Fire*. An excellent reference source for any fire service library and an excellent supplement to **Private Fire Protection.** 1st Edition (1986), 16 pages.

FIRE DEPARTMENT OCCUPATIONAL SAFETY
addresses the basic responsibilities and qualifications for a safety officer and the minimum requirements and procedures for a safety and health program. Included in this manual is an overview of establishing and implementing a safety program, physical fitness and health considerations, safety in training, fire station safety, tool and equipment safety and maintenance, personal protective equipment, en route hazards and response, emergency scene safety, and special hazards. 2nd Edition (1991), 396 pages, addresses NFPA 1500, 1501.

SALVAGE AND OVERHAUL
covers planning salvage operations, equipment selection and care, as well as describing methods and techniques for using salvage equipment to minimize fire damage caused by water, smoke, heat, and debris. The overhaul section includes methods for finding hidden fire, protection of fire cause evidence, safety during overhaul operations, and restoration of property and fire protection systems after a fire. 7th Edition (1985), 225 pages, addresses NFPA 1001.

SELF-CONTAINED BREATHING APPARATUS
contains all the basics of SCBA use, care, testing, and operation. Special attention is given to safety and training. The chapter on Emergency Conditions Breathing has been completely revised to incorporate safer emergency methods that can be used with newer models of SCBA. Also included are appendices describing regulatory agencies and donning and doffing procedures for nine types of SCBA. The manual has been thoroughly updated to cover NFPA, OSHA, ANSI, and NIOSH regulations and standards as they pertain to SCBA. 2nd Edition (1991), 360 pages, addresses NFPA 1001.

STUDY GUIDE FOR SELF-CONTAINED BREATHING APPARATUS
The companion study guide in question and answer format. 1st Edition (1991).

FIRE STREAM PRACTICES
brings you an all new approach to calculating friction loss. This carefully written text covers the physics of fire and water; the characteristics, requirements, and principles of good streams; and fire fighting foams. **Streams** includes formulas for the application of fire fighting hydraulics, as well as actions and reactions created by applying streams under a variety of circumstances. The friction loss equations and answers are included, and review questions are located at the end of each chapter. 7th Edition (1989), 464 pages, addresses NFPA 1001 and NFPA 1002.

GASOLINE TANK TRUCK EMERGENCIES
provides emergency response personnel with background information, general procedures, and response guidelines to be followed when responding to and operating at incidents involving MC-306/DOT 406 cargo tank trucks. Specific topics include: incident management procedures, site safety considerations, methods of product transfer, and vehicle uprighting considerations. 1st Edition (1992), 60 pages, addresses NFPA 472.

FIRE VENTILATION PRACTICES
presents the principles, practices, objectives, and advantages of ventilation. It includes the factors and phases of combustion, flammable liquid characteristics, products of combustion, backdrafts, transmission of heat, and building construction con-

siderations. The manual reflects the new techniques in building construction and their effects on ventilation procedures. Methods and procedures are thoroughly explained with numerous photographs and drawings. The text also includes: vertical (top), horizontal (cross), and forced ventilation; and a glossary. 6th Edition (1980), 131 pages, addresses NFPA 1001.

FIRE SERVICE PRACTICES FOR VOLUNTEER AND SMALL COMMUNITY FIRE DEPARTMENTS
presents those training practices that are most consistent with the activities of smaller fire departments. Consideration is given to the limitations of small community fire department resources. Techniques for performing basic skills are explained, accompanied by detailed illustrations and photographs. 6th Edition (1984), 311 pages.

WATER SUPPLIES FOR FIRE PROTECTION
acquaints you with the principles, requirements, and standards used to provide water for fire fighting. Rural water supplies as well as fixed systems are discussed. Abundant photographs, illustrations, tables, and diagrams make this the most complete text available. It includes requirements for size and carrying capacity of mains, hydrant specifications, maintenance procedures conducted by the fire department, and relevant maps and record keeping procedures. Review questions at the end of each chapter. 4th Edition (1988), 268 pages, addresses NFPA 1001, NFPA 1002, and NFPA 1031, levels I & II.

TEACHING PACKAGES

LEADERSHIP
This teaching package is designed to assist the instructor in teaching leadership and motivational skills. Cause and effect, behavior and consequences, listening and communications are themes throughout the course that stress the reality of the job and the people one deals with daily. Before each lesson is a title page that gives an outline of the subject matter to be covered, the approximate time required to teach the material, the specific learning objectives, and the references for the instructor's preparation. Sources for suggested films and videotapes are included.

CURRICULUM PACKAGE FOR IFSTA COMPANY OFFICER
A competency-based Teaching Package with lesson plans and activities to teach the student the information and skills needed to qualify for the position of Company Officer. Corresponds to **Fire Department Company Officer**, 2nd Edition.

The Package includes the Company Officer Instructor's Guide (the how, what, and when to teach); the Student Guide (a workbook for group instruction or self-study); and 143 full-color overhead transparencies.

ESSENTIALS CURRICULUM PACKAGE
A competency-based teaching package with lesson plans and activities to teach the student the information and skills needed to qualify for the position of Fire Fighter I or II. Corresponds to **Essentials of Fire Fighting**, 3rd Edition.

The Package, with Instructor's Guide, Student Guide, and more than 400 transparencies, is scheduled for publication in 1993.

TRANSLATIONS

LO ESENCIAL EN EL COMBATE DE INCENDIOS
is a direct translation of **Essentials of Fire Fighting**, 2nd edition. Please contact your distributor or FPP for shipping charges to addresses outside U.S. and Canada. 444 pages.

PRACTICAS Y TEORIA PARA BOMBEROS
is a direct translation of **Fire Service Practices for Volunteer and Small Community Fire Departments**, 6th edition. Please contact your distributor or FPP for shipping charges to addresses outside U.S. and Canada. 347 pages.

OTHER ITEMS

TRAINING AIDS
Fire Protection Publications carries a complete line of videos, overhead transparencies, and slides. Call for a current catalog.

NEWSLETTER
The nationally acclaimed and award winning newsletter, *Speaking of Fire*, is published quarterly and available to you free. Call today for your free subscription.

All manuals published by Fire Protection Publications are copyrighted by Oklahoma State University. For further information contact: Fire Protection Publications, IFSTA Headquarters, Oklahoma State University, Stillwater, OK 74078-0118, OR CALL 1-800-654-4055.